T0137791

Ignitability and Explosibility of Gases and Vapors

Tingguang Ma

Ignitability and Explosibility of Gases and Vapors

Tingguang Ma
Fire Protection and Safety Technology
Oklahoma State University
Stillwater, OK
USA

ISBN 978-1-4939-4953-3 ISBN 978-1-4939-2665-7 (eBook)
DOI 10.1007/978-1-4939-2665-7

Springer New York Heidelberg Dordrecht London

Softcover reprint of the hardcover 1st edition 2015

Printed on acid-free paper

Springer Science+Business Media LLC New York is part of Springer Science+Business Media
(www.springer.com)

Preface

Before starting your journey reading this monograph, I want to explain three concepts from etymology perspective (using only etymology dictionary at http://etymonline.com/).

The first concept is "inflame," from which flammability is derived. It first appeared in the mid-fourteenth century, "make (someone) ardent; set (the spirit, etc.) on fire" with a passion or religious virtue, a figurative sense, from Old French enflamer, from Latin inflammare "to set on fire, kindle," figuratively "to rouse, excite," from in- "in" + flammare "to flame," from flamma "a flame." Literal sense of "to cause to burn" first recorded in English in the late fourteenth century. Meaning "to heat, make hot, cause inflammation" is from the 1520s. In order to get a flame, you need fuel/air combination, which is difficult to characterize, since air itself is a mixture of oxygen/nitrogen. The tertiary nature of the mixture makes a universal consistent flammability theory difficult to establish.

The second concept to appear is "explode," "to reject with scorn," from Latin explodere "drive out or off by clapping, hiss off, hoot off," originally theatrical, "to drive an actor off the stage by making noise," hence "drive out, reject, destroy the repute of" (a sense surviving in an exploded theory), from ex- "out" (see ex-) + plaudere "to clap the hands, applaud," which is of uncertain origin. Athenian audiences were highly demonstrative: clapping and shouting approval, stamping, hissing, and hooting for disapproval. The Romans seem to have done likewise. So the act of "explode" has nothing to do with a fuel, simply related to the capability of air to support noise (acoustic wave) or flame (combustion wave). In this monograph, the latter is used for combustion safety, not explosion safety.

Finally, "Ignite" appeared in the 1660s, from Latin ignitus, past participle of ignire "set on fire," from ignis "fire." Igneous is derived from Latin igneus "of fire, fiery," from ignis "fire," or Sanskrit agnih "fire, sacrificial fire." So the latest word has a much remote origin. Perhaps, it is the extreme weather in the 1660s that made the fire initiation process difficult, so an ancient word was revived to describe the difficulties in setting a fire. However, there is no air involved in the ignition process, assuming air is always sufficient.

Based on the original meanings, the explosibility of air and the ignitability of fuel are redefined to describe the flammability of fuel/air mixture. Similar concepts have already been utilized in industry. The explosibility of air is called "In-Service Oxygen Concentration (ISOC)," while the ignitability of fuel is called "Out-of-Service Fuel Concentration (OSFC)." By isolating the concepts of ignitability and explosibility from the concept of flammability, and correlating them with the simple oxygen calorimetry, a new flammability theory is born. It shows mass transfer is more important than heat transfer, though radiative loss is only important in determining the flammability. In order to understand critical behaviors for gases and vapors, we need to understand first principles in heat and mass transfer. This is the starting point of this monograph. Hope you can have a comfortable journey with these concepts.

In addition, I would like to take this opportunity to thank Prof. Steven Spivak for accepting me into the ENFP program at the University of Maryland to study fire. Without his interests in my career goal, it would have been difficult for me to pursue fire studies 16 years ago. My second thanksgiving should be to Prof. Jose Torero, whose teaching on advanced fire suppression technologies in the spring of 2001 lent me the foundation to build this flammability theory. I am also indebted to Prof. Arnaud Trouvé, from whose numerical training I developed an interest in theoretical development. I would also like to thank my colleagues, Drs. Michael Larrañaga and Qingsheng Wang, whose kind help supported my teaching work at OSU. Special thanks should be extended to my former students, Caleb Scheve, Nash McMurtrey, and Kevin Stamper for supporting my writing. They are the targeted audience of this monograph and were willing to help me to get this project through. I am proud of them.

Stillwater, USA Tingguang Ma

Contents

Nomenclature, Subscripts, and Abbreviations

Nomenclature

C_O The oxygen coefficient in a reaction, dimensionless
H_O The heating potential of oxygen based on air, dimensionless
H_F The heating potential of fuel based on air, dimensionless, $H_F = C_O H_O$
Q_D The quenching potential of diluent based on air, dimensionless
Q_F The quenching potential of fuel based on air, dimensionless
Q_O The quenching potential of oxygen based on air, dimensionless
x_L Lower Flammability Limit (volume ratio), %, or dimensionless
x_U Upper Flammability Limit (volume ratio), %, or dimensionless
R Diluent/fuel volumetric ratio
x Concentration of a constituent in the mixture
μ Stoichiometric reaction constants
ν Non-stoichiometric reaction constants
ϕ Equivalence ratio
λ Oxygen concentration
β Fuel concentration
γ Second fuel (doping) concentration

Subscripts

D Diluent-based potential
F Fuel-based potential
L Lower flammable limit
LU The cross point of LFL/UFL lines, or the inertion point
O Oxygen-based potential
U Upper flammable limit
0 Initial state/concentration

Abbreviations

CAFT	Critical Adiabatic Flame Temperature
HQR	Heating–Quenching Ratio
IAR	Minimal inert/oxidizing gas ratio
ICR	Minimal inert/flammable gas ratio
ISOC	In-service Oxygen Concentration
LDC	Limiting Dilution Concentration, $LDC = 1 - \beta_2$
LEL	Lower Explosive Limit
LFC	Limiting Fuel Concentration, β_2
LFL	Lower Flammability Limit
LIC	Limiting Inertion Concentration, $LIC = 1 - \lambda_2$
LIL	Lower Ignitable Limit
LOC	Limiting Oxygen Concentration, λ_2
MAI	Minimum required inert gas concentration
MFC	Maximal Fuel Concentration, β_1
MIC	Minimal Inerting Concentration, $MIC = 1 - 4.773\lambda_1 - \beta_1$
MIR	Minimal Inerting (inert gas/air or oxidizing gas) Ratio, $MIR = (1 - \gamma_1)/\gamma_1$
MMF	Maximum Mixture (diluent + fuel) Fraction, $MMF = 1 - 4.773\lambda_1$
MMR	Minimal Molar (inert/flammable gas) Ratio, $MMR = (1 - \beta_1)/\beta_1$
MOC	Minimal Oxygen Concentration, λ_1
MSDS	Material Safety Data Sheets
MXC	Maximal permissible Flammable gas Concentration
OSFC	Out-of-Service Fuel Concentration
UEL	Upper Explosive Limit
UFL	Upper Flammability Limit
UIL	Upper Ignitable Limit

List of Figures

List of Tables

Chapter 1
A Historical Review

1.1 Dawn of Combustion Science for Safety

Two centuries ago, a mine fire at Felling Colliery near Newcastle shocked Britain, which claimed 92 lives on 25 May 1812. After this disaster, Sir Humphrey Davy was invited to find "a method of lighting the mines from danger, and by indicating the state of the air in the part of mine where inflammable air was disengaged, so as to render the atmosphere explosive, should oblige the mines to retire till the workings were properly cleared [1]." After one year of experimental work, he published his report, releasing his findings on flammability and principles of Davy Safety lamp. Without sufficient instruments (see Fig. 1.1), Davy investigated fire chemistry, flammability and suppressibility, and combustion toxicity of firedamp (mainly methane), which are still insightful in today's views. With modern science on combustion safety, we can have a better understanding of his findings.

1.1.1 *Fire Chemistry*

"One measure of it required for its complete combustion by the electric spark nearly 2 measures of oxygen, and they formed nearly 1 measure of carbonic acid." This is to say, the combustion chemistry has a stoichiometric oxygen/fuel molar ratio of two. This is consistent with $C_O = 1 + 4/4 = 2$ for methane (CH_4). We still use this principle in thermochemistry or combustion science, where a global energy balance is necessary to understand the energy releasing process.

T. Ma, *Ignitability and Explosibility of Gases and Vapors*,
DOI 10.1007/978-1-4939-2665-7_1

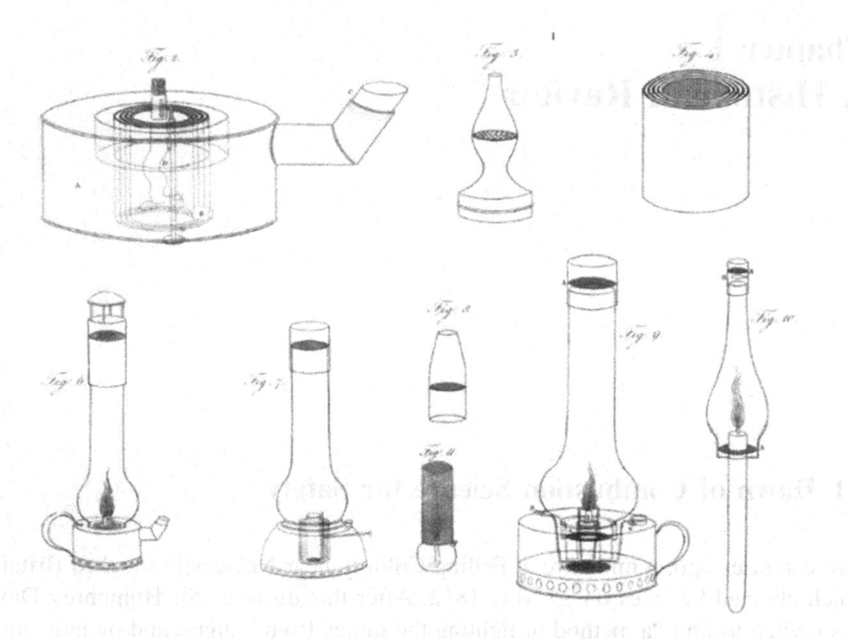

Fig. 1.1 Sir Humphrey Davy's experimental tools for flammability testing [1]

1.1.2 Flammability

"In 1 part of gas and 15 parts of air, the candle burnt without explosion with a greatly enlarged flame; and the effect of enlarging the flame, but in a gradually diminishing ratio, was produced as far as 30 parts of air to 1 of gas." So his lower flammability limit was estimated as 0.5*(1/16 + 1/31) = 0.0474.

"… a common electrical spark would not explode 5 parts of air and 1 of fire-damp, though it explode 6 parts of air and 1 of damp". This translates to an upper flammability limit of 0.5*(1/6 + 1/7) = 0.155.

Now we know that the official MSDS flammable range for methane is 5–15 %, not far away from his observations.

1.1.3 Suppressibility

"Azote, when mixed in the proportion of 1–6 of an explosive mixture, containing 12 of air and 1 of fire-damp, deprived it of its power of explosion". The original mixture has a molar fuel concentration of 1/13 = 7.7 %, which falls within the flammable range of 5–15 %. With azote (old name of nitrogen) dilution, the fuel concentration is 7.7 % * (6/7) = 6.6 %, still flammable, but his ignition source (a spark) cannot ignite this mixture.

"1 part of carbonic acid to 7 of an explosive mixture deprived it of the power of exploding; so that its effects are more remarkable than those of azote." Since less amount of carbonic acid (carbon dioxide) is used for dilution, so he concluded that carbonic acid had a greater capacity of heat or a higher conducting power due to its greater density. This is consistent with our observation that carbon dioxide has a quenching potential 1.75 times of air, while nitrogen is only 0.992 times of air.

1.1.4 Detection by Upper Limits

"When the fire-damp is so mixed with the external atmosphere as to render it explosive, the light in the safe lantern or lamp will be extinguished, and warning will be given to the miners to withdraw from, and to ventilate, that part of the mine." This is a special feature about Davy's safety lamp. Since the gauze allow fuel and air free to migrate, high concentrations of fuel gases may render the mixture non-explosive (short of oxygen), depriving the mixture of its explosibility (oxygen), so the lamp fire is quenched, giving out a sufficient warning of a gas leak. However, since this practice requires a dangerous operation to cross the flammable zone, this principle is seldom recommended for practical uses.

1.1.5 Minimum Ignition Energy

"When explosions occur from the sparks from the steel mill, the mixture of the fire-damp is in the proportion required to consume all the oxygen of the air, for it is only in about this proportion that explosive mixtures can be fired by electrical sparks from a common machine". Now we know, the minimum ignition energy happens near a stoichiometric composition, slightly tilted over the rich side (see Fig. 1.2). The ignition threshold is smaller, if the mixture is close to its stoichiometric composition. From the flame temperature theory, the stoichiometry means there is no extra fuel and no extra oxygen for heat absorption, so the required energy can be smaller.

1.1.6 Combustion Toxicity

"Supposing 1 of fire-damp to 13 of air to be exploded, there will still remain nearly 1/3 of the original quantity of oxygen in the residual gas." Applying the chemical balance to the combustion of firedamp, we have

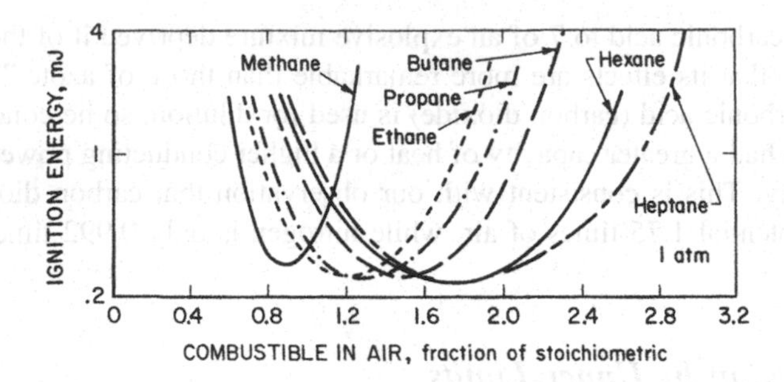

Fig. 1.2 Spark ignition energy versus combustible vapor concentration for six paraffins in air at atmospheric pressure [2]

$$CH_4 + \mu_{air} \cdot (O_2 + 3.773 \cdot N_2) = \mu_{CO_2} CO_2 + \mu_{H_2O} \cdot H_2O + \mu_{O_2} \cdot O_2 + 3.773\mu_{air} \cdot N_2$$

Since the initial oxygen/fuel molar ratio is $\mu_{air} = 13/4.773 = 2.72$, we have

$$CH_4 + 2.72 \cdot (O_2 + 3.773 \cdot N_2) = CO_2 + 2 \cdot H_2O + 0.72 \cdot O_2 + 10.3 \cdot N_2$$

Therefore, the remaining oxygen is 0.72/2.72 = 26.5 %, or 1/3 of the original quantity of oxygen in the residual gas. Here the oxygen molar concentration after explosion is 0.72/(1 + 2 + 0.72 + 10.3) = 7.1 %. Davy compared this value with his extreme case, "an animal lived, though with suffering, in a gas containing 100 parts of azote, 14 parts of carbonic acid, and 7 parts of oxygen." This translates to a level of 5.8 % oxygen. He attempted to understand whether the victims could survive the oxygen-depleted environment after gas explosion. At current level of understanding on human toxicity, this problem is still far from being solved.

Though primitive, Davy pioneered the scientific research on safety. After 200 years of experimental explorations, his method of ignition (spark ignition), his combustion chemistry, his criteria of flame propagation, and a few other principles in performing flammability tests, are still in use today. What is improved is the standardization of equipment and better instrumentation [3]. What is still missing is a fundamental and unifying theory behind numerous experimental efforts. To supply a fundamental theory is the primary purpose of this monograph.

1.2 Continued Efforts After Davy

1.2.1 Three Types of Flammability

Nearly 140 years after Davy's work, it was still claimed that "nobody has succeeded in calculating either a lower or a higher limit of flammability of any mixture from more fundamental physicochemical data" [4]. Even now with latest computers, the flammability problem is still solved case by case, without a coherent and fundamental theory to support engineering applications. Le Chatelier's rule (first proposed in 1891) is still in use today for mixture flammability, while the modified Burgess-Wheeler's law (first proposed in 1911) is also still in a dominant position for variable ambient temperature. Why a reasonable flammability theory is so difficult to establish?

As Egerton [5] states, an "inflammability limit" or the **flammability**, is so defined that flame can be propagated indefinitely if the composition of the mixture and the conditions are the same. Other researchers used "self-propagation" [6], "self-support" [7], "self-sustained and propagate" [8] to refine the threshold "propagate indefinitely". Such a criterion on flame propagation is chosen based on following considerations.

- Scale-invariance. As Sir Alfred Egerton [5] believed, for practical purposes it is important that the limits should be determined under those conditions which represent a good approximation to the limits obtained in large volumes of gas mixtures.
- Ignition strength. As Burgess and Wheeler [6] stated, the ignition source should be strong enough to raise the gas mixture to its ignition-temperature, while the flame is propagated without the influence from the heat source.
- Boundary impact. Britton [3] reviewed 200 years of experimental work, the boundary impact is a major factor on selecting the right platform. It took nearly 150 years to find the Bureau of Mines apparatus as the standard platform for flammability testing.

The characterizing feature about a **flammability** test is the flame propagation distance, which should be halfway of the flammable space. If the fuel composition supports the flame sweeping throughout the space, it is called explosibility. If the flame is just ignited, but cannot propagate away from the ignition source, it is called ignitability. These are three parallel concepts on mixture safety, however, only flammability gains a popular position, while the other two cannot be tested and theoretically justified, so improperly defined.

Here, **ignitability** is defined as the critical fuel concentration that does not support ignition initiation. Ignitability does not have any dependence on oxygen or environmental factors, so it is fundamental to that fuel. However, unlike a visible flame front moving in flammability tests, the ignition criteria in ignition tests are difficult to establish. This is a limiting case, physically meaningful, but difficult to test or observe.

Explosibility is defined as the ability for a critical flame to propagate throughout the entire volume of the mixture and develop considerable pressure [9], while flammability is used to describe those limiting mixtures within which flame will propagate through the mixture indefinitely, irrespective of whether or not pressure is developed. Generally, explosibility requires a strong ignition source and a higher-pressure criterion. Since the oxygen in background air supports flame propagation, explosibility is a property of background air, or specifically of oxygen. Since the full reaction status is difficult to quantify and validate, pressure criterion is proposed for testing explosibility. However, in order to match flammability data, different pressure criteria (2, 3 and 7 %) are adopted in different experimental efforts. Since both explosibility and flammability are limited by oxygen, and they have a small difference between each other, they were assumed to be interchangeable [9].

Based on these definitions, we have three limiting concentrations, Lower Ignition Limit (LIL), Lower Flammability Limit (LFL) and Lower Explosion Limit (LEL). They are qualitatively demonstrated in an Arrhenius-type temperature curve shown in Fig. 1.3. Theoretically, if a smaller ignition criterion is chosen, the tested limit is ignitability, which is a property of the fuel. If a higher-pressure criterion is chosen, the tested limit is explosibility, which is a property of background air (or available oxygen). Flammability is a process property, which lies in between these two fundamental properties. This leads to the difficulty and complexity on establishing a material property database for flammability, which is more process-dependent. Table 1.1 lists three concepts and their major differences. More theoretical derivations and physical meanings will be provided in Chap. 5.

The difficulty in defining these three concepts is that there is no satisfactory definition developed for the word “flame” or “ignition”. In most cases, flames involve strongly exothermic reactions between gases or vapors resulting in hot combustion products, usually at a temperature above 1400 K and accompanied by light emission, under confinement, flames typically lead to sharp rises in pressure. It is through the criteria on temperature, radiation and pressure that we judges whether an ignition has occurred. Van Dolah et al. [10] discussed the impact of various ignition sources on flammability.

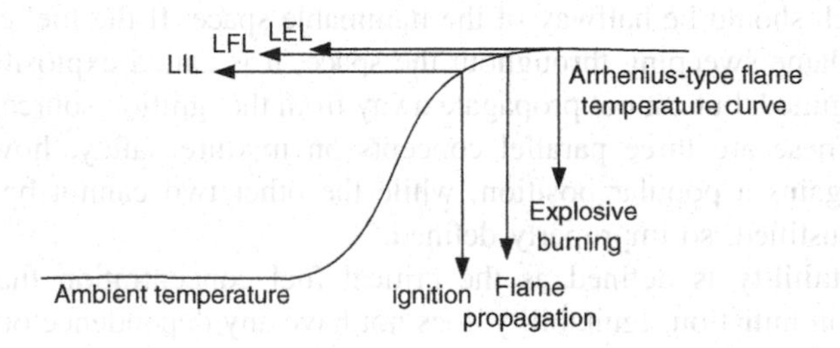

Fig. 1.3 The Arrhenius-type temperature curve showing the difference between LIL/LFL/LEL

Table 1.1 Critical concepts relating to flammability

Mixture	Definition	Sub concepts	Controlling factor
Explosibility	The mixture is ignitable, support flame propagation throughout the mixture	LEL/UEL, LOC/LIC	Air (oxygen)
Ignitability	The mixture is ignitable if available oxygen is just sufficient to start reaction/ignition	LIL/UIL, LFC/LDC	Fuel
Flammability	The mixture is ignitable, support flame propagation halfway	LFL/UFL, MOC/MFC	Fuel/air (oxygen)

For European flammability standards using pressure criteria, if a LFL measurement is taken under a lower pressure threshold, it is approaching the lower ignition limit (LIL). If a higher pressure threshold is taken as the ignition criterion, the sample is assumed to have a sweeping flame established, or the lower explosive limit (LEL). Depending on the pressure criteria used, European flammability standards can produce ignitability, flammability and explosibility. However, under most circumstances their flame propagation criteria are higher than a comparable "flammability" test, they generally report explosibility data rather than flammability limits. So their data have a built-in safety factor [11]. This is the reason that the experimental data from European standards are generally more conservative than US data. Britton [3] recommended that flammable limits measured in European standards, such as DIN51649 or prEN1839, should not be mixed with other database, since the flame propagation criteria are not fully met in these standards.

Note, flammability and ignitability have different meanings in terms of liquid safety [12]. The light hydrocarbon fuels can produce flammable mixtures at near ambient temperatures and thus for fire safety require protection from sparks, flames and other localized sources of energy within the vicinity of storage. The heavier fuels, including lubricants, are not flammable in this way, but would ignite spontaneously if subjected to general overheating. Therefore, the ambient temperature decides the liquid fuel to be flammable or ignitable. However, this is limited to liquid safety only.

Since the testing theory for flammability is far lagging behind diverse new materials and applications, there is a need for clarifying basic concepts and providing a new theory to check the safety concerns systematically. This monograph will serve this role to review previous experience and solve current engineering problems from fundamental principles.

1.2.2 Experimental Efforts

In the century after Davy, several researchers tested the flammability of methane with improved resolution, while their experimental setups were questionable in today's view. The debating focus is always the ignition strength and boundary heat losses. In order to avoid the heat loss through the boundary, a spherical vessel was proposed to isolate the wall impact (see Fig. 1.4). The flame spread criterion is difficult to establish in such a device, so a pressure criterion is proposed instead. This explosibility testing device was soon given up in favor of the flame propagating tube for flammability testing. However, in 1980s, there was a revival of interests on spherical vessels, mostly undertaken by European researchers [3].

In the early half of the 20th century, Bureau of Mines took the lead in harmonizing the experimental work on flammability. Under a belief that flammability and explosibility are interchangeable, American experimental efforts are focusing on flammability tests, with numerous improvements on controlling heat losses in various forms. Jones [9] summarized factors affecting the limits of flammability, including the direction of flame propagation, the design, diameter, and length of the

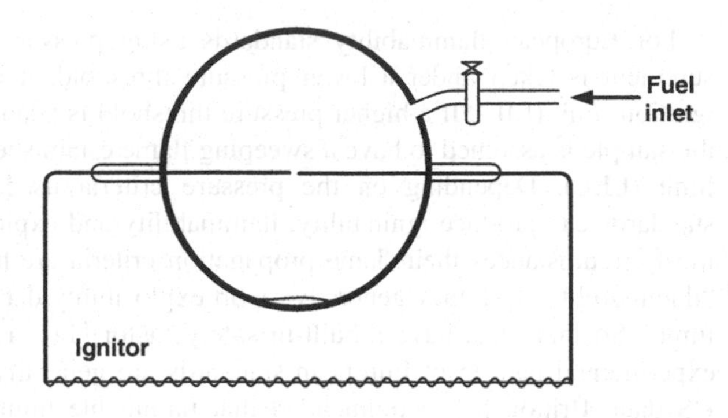

Fig. 1.4 Typical explosibility tester by Burgess et al. [6]

test apparatus, the temperature and pressure of the mixture at the time of ignition, the percentage of water vapor present, and indirectly by the source of ignition. Since he stressed so many heat loss terms, his definition is really flammability, which is sensitive to various heat losses.

Currently, the most known apparatus for determining the flammability of a premixed gas was developed by Bureau of Mines. It consists mainly of a vertical tube, with a flame propagating from the ignition source in the premixed gases (shown in Fig. 1.5). Most fuels are tested in gaseous or vapor form, with a conditioning device for a controlled vapor production. After premixing with air for 10–30 min, the premixed gases was ignited from the lower end of the tube. If the flame propagates at least halfway (75 cm) up the tube, the mixture is declared flammable. Otherwise, the test has to be restarted at a different concentration after purging and reconditioning. Due to the presence of buoyancy, the upward flame spread is easier than downward flame spread, so the upward propagation usually generates a wider flammability limits, conservative for safety reasons. Thus, lower-end ignition is adopted for producing flammability limits in all tabulations and calculations. Though this apparatus is cumbersome and no longer popular, most published flammability data were collected on this platform.

After the BOM flammability apparatus, new apparatus are proposed, including glass sphere (ASTM E681) and ISO tests, with a continued effort to minimize the wall impact. These tests are simpler to perform, but with the returned drawbacks of ignition-dependence and unsatisfying flame propagation [3].

Note Simmons and Wolfhard [114] and Ishizuka and Tsuji [113] used a diffusion flame for measuring ignitability and explosibility, though they did not realize their conceptual difference from the flammability. A typical setup is shown in Fig. 1.6.

After 200 years of experimental work, all tests are generally classified into two categories, flammability test and explosibility test. The former stresses the independence of ignition source, while the later stresses the isolation of boundary heat loss. Generally, explosibility tests are performed in an explosion ball/sphere with central ignition, while flammability tests are performed in an elongated tube with bottom ignition. Without an agreement on controlling various heat loss terms, no flammability test has gained wide acceptance on precision and reproducibility as the standard Bureau of Mines apparatus.

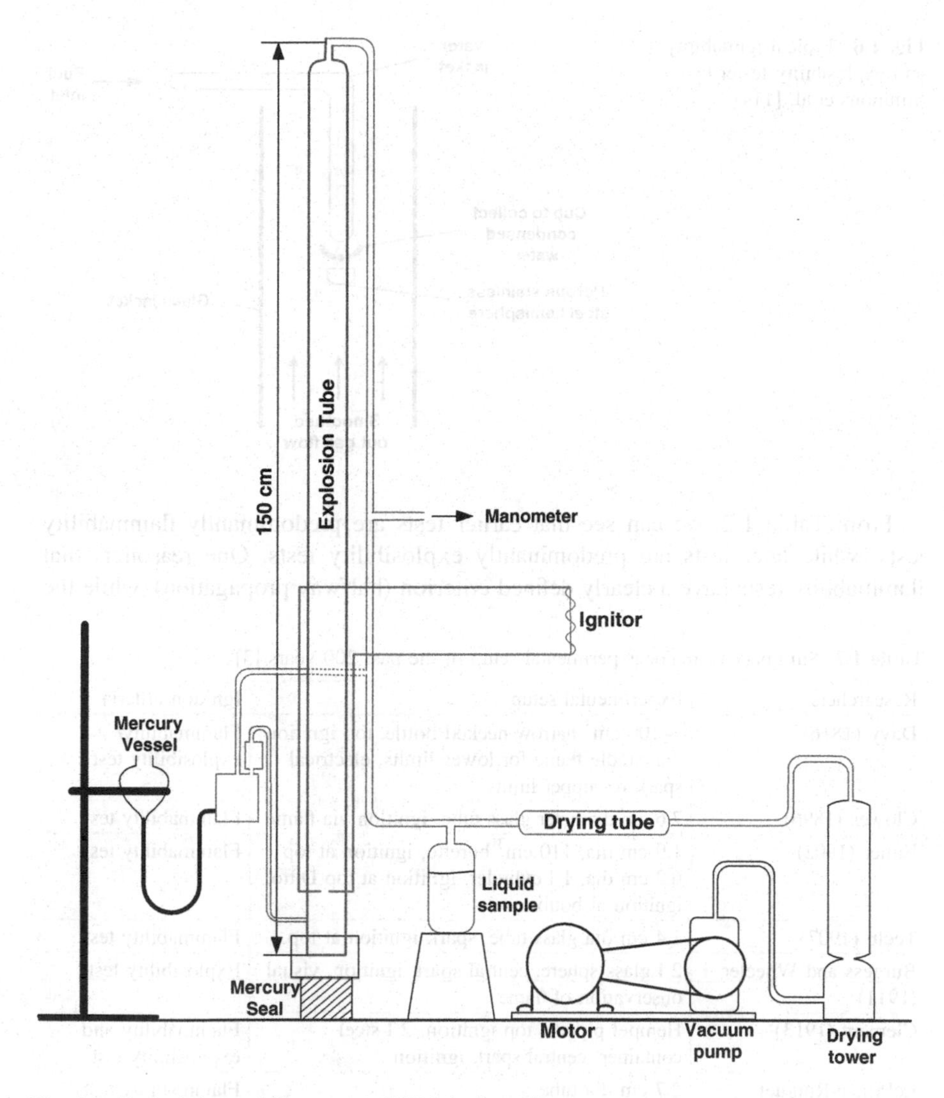

Fig. 1.5 Typical flammability testing BOM devices [4]

The experimental history of flammability measurement is composed of the competition between flammability and explosibility tests, as shown in Table 1.1. Britton [3] reviewed various experimental methods to characterizing the flammability, including the famous candle experiment by Sir Humphrey Davy in 1816. Most of previous works focused on keeping the correct reaction kinetics by removing the sources of disturbance such as the heat losses and the impact of ignition source strength. There is no consensus on which criterion is better, since a fundamental theory is missing.

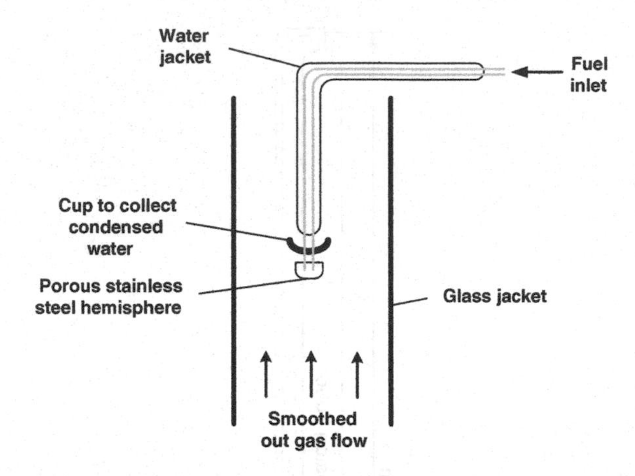

Fig. 1.6 Typical ignitability and explosibility tester by Simmons et al. [114]

From Table 1.2, we can see that earlier tests are predominantly flammability tests, while later tests are predominantly explosibility tests. One reason is that flammability tests have a clearly defined criterion (halfway propagation), while the

Table 1.2 Summary of major experimental setup in the past 200 years [3]

Researchers	Experimental setup	Ignition criteria
Davy (1816)	~100 cm^3 narrow necked bottle, top ignition via candle flame for lower limits, electrical spark for upper limits	Flammability/ explosibility test
Clowes (1896)	7.6 cm diameter glass tube, ignition via flame	Flammability test
Eitner (1902)	1.9 cm dia, 110 cm^3 burette, ignition at top 6.2 cm dia, 1 l cylinder, ignition at top Ditto, ignition at bottom	Flammability test
Teclu (1907)	1.4 cm dia glass tube, spark ignition at top	Flammability test
Burgess and Wheeler (1911)	2 l glass sphere, central spark ignition, visual observation of flame	Explosibility test
Clement (1913)	Hempel pipette, top ignition, 2 l steel container, central spark ignition	Flammability and explosibility test
LePrince-Ringuet (1914)	2.7 cm dia tube	Flammability test
Coward and Brinsley	2.7 cm dia tube 11 l bottle, complete propagation with top ignition Vertical 170 l box, 30 cm square section × 1.8 m long, standing over water	Flammability test
Burrell and Oberfell (1915)	163 l vertical box, upward propagation	Flammability test
Jones et al. (1933)	10 cm × 96 cm vertical pipe, 41 % pressure rise criterion	Flammability test
Simmons and Wolfhard (1955)	A counterflow diffusion flame for limit oxygen index	Explosibility and ignitability test

(continued)

Table 1.2 (continued)

Researchers	Experimental setup	Ignition criteria
Christner (1974)	7 l steel cylinder, various ignition strength	Explosibility test
Ishizuka and Tsuji (1981)	A counterflow diffusion flame established in the forward stagnation region of a porous cylinder	Explosibility and ignitability test
Burgess et al. (1982)	25.5 m^3 sphere: visual and photographic observation	Explosibility test
Hertzberg and Cashdollar (1983)	7 % total pressure rise criterion for 8 l chamber, 20 l chamber, 120 l sphere	Explosibility test
Mashuga and Crowl (1998)	20 l sphere: 10 J central fuse wire igniter, 7 % pressure rise criterion	Explosibility test
De Smedt et al. (1999)	DIN 51649, 20 l sphere	Explosibility test
Cashdollar et al. (2000)	20 l chamber, 120 l sphere, 3 % pressure rise criterion	Explosibility test

pressure criterion for explosibility tests are not widely accepted. From the theories in this monograph, explosibility and ignitability are ideal and fundamental concepts, which require limiting conditions difficult to satisfy directly. They can only be extrapolated from a flammability diagram, giving the flammability diagram a unique position to derive material properties (see Chap. 5).

1.3 Who Still Cares About Flammability?

Different times saw different needs from the flammability theory. During the time prior to Coward's BOM report, the academic interests are primarily on **mine safety**, as Fieldner said [4], "a knowledge of limits of flammability of methane and of the distillation products of coal in air and in partly vitiated atmosphere is of fundamental importance in the study and prevention of mine explosions". Even with modern instrumentation and management experience, flammable gas fires are still imposing a serious problem, such as the coalmine fire in Turkey (May 13, 2014).

Then as the petrochemical industry boomed, **chemical and process safety** gained widespread attentions on synthetic chemicals. This trend [4] could be identified from the number and diverse requests to the Bureau of Mines for information on the limits of flammability of various gases and vapors when mixed with air and other "atmosphere". The explosion of apartment buildings in East Harlem (March 13, 2014) is a typical example of fire safety problem of a flammable mixture in society.

In 1990s, with the political decision to phase-out Halon, numerous candidate refrigerants were proposed for replacing Halons. Some agents are flammable, thereby introducing the safety concerns during accidental discharge of refrigerants.

How to control the hazards of explosive burning of a flammable leak? How to design a refrigerant mixture with an optimum performance without introducing additional danger of ignition and fire? In addition, most flammable mixtures in commercial market have a potential problem of explosive burning, or explosion, thereby demanding a comprehensive evaluation on their ignition potentials for **refrigerant safety**.

On the other hand, numerous agents were proposed in the suppression field about the same time. Material properties have been sifted thoroughly to find the best agent without success. A good candidate with a comparable performance of Halon 1301 can never be found. Attention has been shifted to the design of suppressant mixtures. How to design a suppressing mixture, which provides the best chemical synergy without endangering the environment? The only hope is to check the material properties and optimize the performance of a mixture. One motivation behind flammability research comes from **suppression engineering**.

In the new millennium, under the pressure for environmental-friendliness and fuel economy, clean combustion technologies became a hot research field, while the flammability problem (or more precisely the ignitability problem) is of fundamental importance to the safe operation of new combustion technologies, or specifically, **combustion safety**.

1.4 How This Methodology Is Developed?

Rome was not built in one day. My first work on liquid safety is an internship with Combustion Science Engineering, Inc. with experiments on liquid burning on carpet, through which it was found that the wick effect is controlled by heat transfer rather than mass transfer, or the heat balance is more critical for evaporation [13]. It is a simple application of fundamental principles in energy conservation, which gave me some insights on all critical behaviors.

My interests on suppression started with Dr. Jose Torero's class on advanced fire suppression theory in the spring of 2001. Because of the challenging nature and lacking a consistent theory, I was fascinated with suppression theory, especially when my first rotation position upon graduation was at Kidde-Fenwal, Inc. the largest manufacturer for clean agent suppression systems. Then I read widely those papers in HOTWC proceedings, with a macroscopic view on various suppression mechanisms. In order to interpret the suppression behavior, my initial reaction was to use flammability data, treating air as a diluent. This strategy leads to a thermal view on suppression [14] and a thermal balance method for flammability [15]. The reaction chemistry is almost completely bypassed as the synergistic effect was covered as a result of the raised flame temperature threshold.

Since the critical suppression concentrations can be explained thermally, attention was shifted to interpret flammability data with a simple method. By converting the concentration into energy terms (thermal signature) which are additive, the thermal balance method is equivalent to Le Chatelier's rule and solves the cases which the later cannot do, such as the role of oxygen, temperature, diluent, fuel, etc. This method has been applied to flame temperature [16], correlations [17], coal mine fires [18], oxy-combustion [19], etc. It is not very precise to match the experimental data, but sufficient to meet requirements for most engineering applications.

At the suggestions of Mr. Samuel Roger, I adopted the concept of diluent/fuel ratio to derive dilution requirements. Later the governing equations were also found out to be the theoretical boundaries for flammable envelopes [20]. Thus the flammability diagram of all pure fuels can be reconstructed, with known system errors. Then analytical solutions for dilution and purge operations within theoretical flammability diagrams [21, 22] were developed. Following the advice of Mr. David Willson from Stanbridge Capital, I solvded some flammability problems in burning low-calorific-value gases, which formed the basis of type II problem. Here the most important discovery is the concepts of ignitability, flammability and explosibility, which lay the foundation for a consistent engineering methodology and systematically solve most gas-related safety problems. If all critical lines can be analytically defined, the utility of any flammability diagram is also improved.

This monograph is devoted to solving generic and engineering flammability problems from fundamental principles. First, a review of classical flammability theories in history is provided in Chap. 2. Then, from the thermochemistry theory for combustion safety (Chap. 3), the thermal balance method is reviewed to derive the critical limits and flammability diagrams from the thermal signature of a fuel (Chap. 4). The most important concepts are proposed and discussed in Chap. 5, which covers two types of problems (type I and II) and three important concepts (ignitability, flammability and explosibility), including different applications of various flammability diagrams. Type I problem deals the flammability in a confined space, where the oxygen is controlling the flammability and explosibility. Type II problem deals the ignitability of a fuel stream releasing into the open air, where the oxygen supply is infinite. With a quantitative analysis of the thermal balance at critical limits, all operations within flammability diagrams are reviewed in Chap. 6. These flammability diagrams are fundamentally equivalent, while they have different priorities in presenting data. Chapter 7 reviews type II problems, including the roles of diluent/fuel/oxygen/temperature on flammability. They are useful for clean-combustion technologies and for inertising an accidental release of a flammable mixture. Compartment fires are typical type I problem, which are reviewed in Chap. 8. Critical behaviors, backdraft and blow-torch effect, are also reviewed within flammability diagrams. Finally, there is a summarizing chapter on the limitations of this method, and a comparison of critical concepts. It will enforce our impression that ignitability and explosibility are fundamental concepts, whereas the flammability of a mixture is a delicate balance between the ignitability of air and the explosibility of fuel.

Chapter 2
Classical Flammability Theories

With the expectation of establishing a fundamental flammability theory, the primary factor that determines the flammability limit is the competition between the rate of heat generation, which is controlled by the rate of reaction and the heat of reaction for the limit mixture, and the external rate of heat loss by the flame [7]. Many theoretical developments are performed to determine the controlling factor, which may be the heat loss and some aerodynamic effects. Here are a few efforts toward this goal.

1. Daniell [23] analytically established a minimum tube radius for flame propagation by analyzing the heat loss through the wall.
2. Flammability limits are explained as the instability to small disturbance from the steady state [24].
3. Flames are governed by a multiple eigenvalue problem [25, 26].
4. Flammability limits are explained in terms of kinetic properties of the mixture [27].
5. Limit appears at the composition at which the flame becomes unstable to a change in the curvature or extent or form of the flame front (Convectional effects [28]).
6. Spalding [29], Mayer [30] and Berlad and Yang [31] have developed an unidimensional flame theory and implied that the radiation loss produces a fundamental limit.
7. According to Lewis and von Elbe [32], the flame propagation in a diverging flow results in that the burned gas moves parallel to the flame and this the cause of convection-induced heat loss from the flame into the cold mixture.
8. Heat loss and chain termination are simultaneously important at these limits [33].
9. The extinction limit occurs as a result of flame temperature reduction when the rate of radiative loss becomes substantial compared with the rate of combustion heat release [34].

They all stress a certain feature interrupting the flame propagation. However, the major mechanism for heat transfer, mass transfer, is not given sufficient attention. Only when the flame temperature has dropped near its critical threshold, the above

T. Ma, *Ignitability and Explosibility of Gases and Vapors*,
DOI 10.1007/978-1-4939-2665-7_2

mechanisms began to play a role. In addition, there are also several conflicting views on the controlling factors. For example, radiation losses represent a small proportion of the total losses, which has been shown by Egerton and Powerling [35] using silvered and blackened tubes. Linnett [28] proposed that the limits obtained in tubes (even the widest ones) are not fundamental. Since there are so many factors controlling the flame propagation process, a consistent flammability theory is difficult to establish. **This is consistent with the fact that flammability is not fundamental to that fuel**. This fact led many researchers to propose empirical rules to manipulate flammability. These empirical rules work for some problems, without the need to know the fundamental difference between flammability and explosibility.

2.1 Empirical Rules on Flammability

2.1.1 Milestone Events on Flammability Theory

Since the pioneering experimental work by Davy in 1816, the first major theoretical breakthrough came with Le Chatelier's empirical rule, which first appeared in 1891 [36]. During a systematic study on diluents and flame temperature, Coward et al. [37] first applied Le Chatelier's rule in its present form. Burgess-Wheeler [6] tried to establish a thermal balance for flammability limits with the concept of flame temperature. The comprehensive flammability data were reported in a series of BOM Bulletins, the most famous one is the last version [4], which appeared in 1952. In 1965, Zabetakis [43] compiled and published latest data for flammability limits, auto-ignition, and burning-rate data for more than 200 combustible gases and vapors in air and other oxidants, as well as of empirical rules and graphs that can be used to predict similar data for thousands of other combustibles under a variety of environmental conditions. His report is the bible on flammability, still in use today. His work remains the core document for supplying flammability information, showing the fact that there is little progress on flammability theory after him. In a series of reports (first appeared in 1975), Hertzberg [38–40] tried to establish an in-depth view on flammability theory, however, his theory is far from engineering applications and left small impact in a computer age. With the development of computer technologies, a detail chemical analysis has been applied for predicting the flammability behavior [41], however, engineering applications are still relying heavily on the empirical Le Chatelier's rule or modified Burgess-Wheeler's law for estimations. Without a fundamental theory, it is difficult to provide an overview of flammability theories, especially when the testing principles behind data are also not well received.

2.1.2 Energy Dependence

It was long recognized that the calorific value of the paraffin is the sole determining factor in calculating the relative lower limits of inflammation of mixture of each with air. Le Chatelier and Boudouard [42] observed that the heat of combustion per unit volume of limit mixtures with air for alkanes was approximately a constant at lower limits, an empirical finding later confirmed by numerous researchers. Out of the simple energy balance at lower flammability limit, Burgess and Wheeler [6] further proposed the principles in deriving the temperature dependence of flammability limits. For the first time, they suggested "the heat liberated by a mole of a lean limit mixture is nearly constant for many combustible-air-mixture at ordinary temperature and pressure", out of which developed the Law of Burgess and Wheeler [43]. That is, the calorific values of the pure paraffin hydrocarbons times their lower limits of inflammability were a constant and that a lower-limit mixture of any of the paraffin hydrocarbons with air on combustion liberates the same amount of heat. White [44] confirmed that the lower limit for downward propagation of flame was approximately inversely proportional to the net calorific value of the vapor used. Jones [9] further concluded that the primary factor that determines the flammability limit is the competition between the rate of heat generation, which is controlled by the rate of reaction and the heat of reaction for the limit mixture, and the external rate of heat flow by the flame. However, such an energy balance was not established and applied properly, probably due to the ternary nature of a mixture.

Based on the observation that the amount of heat contained in the products of combustion of any given layer is just sufficient to raise to its ignition-temperature for the layer adjacent, Burgess and Wheeler [6] established

$$x_L \cdot \Delta H_C = K \tag{2.1}$$

where x_L is the percentage of combustible in the low limit mixture, ΔH_C is the net heat of combustion of the fuel and roughly as $\sim$10.5 kCal/mol. Most existing correlations follow this format to estimate the mixture flammability [17]. It is called "K-constant" method [45], which holds for paraffin up to pentane. Note the original Burgess Wheeler's law [6] used $K = 1059$ for methane. Spakowski [46] used $K = 1040$ to make his simplification. Hanley [47] proposed $K = 1120$, while Ramiro et al. [17] proposed $K = 1042$. They all belong to the "K-constant" method. The scattering of data shows the uncertainty in flammability measurement.

However, earlier researchers, including Burgess and Wheeler, noticed that K is not a constant for some fuels and more heat is required to maintain the combustion of the higher hydrocarbons. The value K increases as the paraffin series is ascended and there is therefore an increase in flame temperature for Cp shows only a relatively small change. The values increase asymptotically to about 1600 K where the effect of dissociation begins to become appreciable [5]. Britton [48] later concludes,

"K Constant" method implies a constant flame temperature at the LFL. White [44] and Zabetakis et al. [49] further proposed

$$x_L \cdot \Delta H_C + \Delta H = K \tag{2.2}$$

where ΔH is the fraction of energy absorbed by ambient air.

Along with the thermal theory, a chain reaction was proposed to explain the thermal phenomena on flame propagation ([35, 50]). Out of the chain reaction theory, it was proposed that some species have more radicals produced in a flame. However, these promoters appeared in the gas ahead of the flame have little specific influence on propagation limits. It is further concluded that the main criterion for inflammation is the maintenance of a sufficient rate of reaction and heat release in the flame, providing thereby sufficient active radicals to inflame the entering gases.

By examining the "K Constant" method, three important outcomes are proposed [48], the modified Burgess-Wheeler law for estimating the temperature dependence, the "Jones' Rule" or "Lloyd's Rule" for predicting LFLs of paraffin hydrocarbons, and most important of all, the Le Chatelier Rule for estimating the flammable limits of mixtures of fuels whose individual LFLs are known. This method is the central principle guiding the experimental and theoretical work for the past 100 years. Most of the "hot wire" combustible-gas indicators operate on this principle directly [48].

2.1.3 Temperature Dependence

The most important parameter affecting the flammability limits is temperature, since an increase in initial temperature leads to a widening of the flammable range [43].

Experimentally, White [44] showed that both limits are affected fairly uniformly. There are several researches on the temperature dependence, using hydrogen [51], low temperature of methane [52], toluene [53], upper explosion limit of lower alkanes and alkenes [54], mixtures of gases [55], NH_3–H_2–N_2–air mixtures [56], pure liquids [57], etc. Their experiments have confirmed that the flammability limits and the initial temperature are related linearly. An important error in experiments can be due to the slow oxidation in the heated vessel before it starts burning.

Rowley et al. [58] summarized the following methods for estimating temperature dependence. Out of the assumption on $x_L \cdot \Delta H_C = K$, the original Burgess-Wheeler Law was expressed as

$$x_L \cdot \Delta H_C = 100 \cdot c_p \cdot T \tag{2.3}$$

where c_p is the specific heat at constant volume of air.

Later the temperature difference was expressed as the enthalpy ratio.

$$\frac{x_L(T)}{x_L(T_0)} = 1 - \frac{\bar{c}_{p,fuel-air-mixture}}{x_L(T_0)\cdot(-\Delta H_C)}(T - T_0) \tag{2.4}$$

Zabetakis [43] further simplified this dependence into a simple constant linear dependence, which is widely adopted in any safety-related textbooks.

$$\frac{x_L}{x_{L,0}} = 1 - \frac{0.75}{x_{L,0}\cdot\Delta H_C}\cdot(T - T_0) = 1 - 0.000721\cdot(T - T_0) \tag{2.5}$$

The second method is proposed by Britton and Frurip [59], which simplifies the temperature dependence as a linear temperature dependence

$$\frac{x_L(T)}{x_L(T_0)} = \frac{T_{AFT} - T}{T_{AFT} - T_0} = 1 - \frac{T - T_0}{T_{AFT} - T_0} \tag{2.6}$$

where T_{AFT} is the adiabatic flame temperature for ignition and flame spread to take place.

Finally, Catoire and Naudet [60] developed an empirical correlation with a simple temperature dependence as

$$x_L(T) = 519.957\cdot X^{0.70936}\cdot n_C^{-0.197}\cdot T^{-0.51536} \tag{2.7}$$

However, they are all related to the energy balance for the background air (temperature or enthalpy). Among them, the modified Burgess-Wheeler Law is well known and widely used in estimating temperature-modified limits. A comparison with experimental data shows there is still some discrepancy between theory and reality (Fig. 2.1). The missing part is the contribution of the fuel, though small, generally ignored in most correlations.

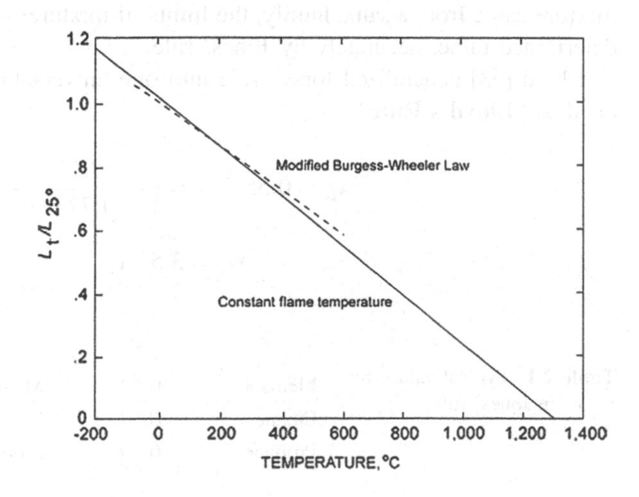

Fig. 2.1 Effect of temperature on L_t/L_{25} ratio of paraffin hydrocarbons in air at atmospheric pressure [43]

Empirically, the temperature dependence is assumed be linear and expressed as the percentage drop per 100 K temperature drop. Zabetakis et al. [49] have found that the lower limits of hydrocarbons decrease linearly by about 8 % for 100 °C rise in temperature. This is consistent with the decrease of net enthalpy rise of background air. Using the propane data of Kondo et al. [61], the experimental decrease per 100 °C is found to be 8.64 %.

This linear temperature dependence applies only if the reaction is complete, the flame temperature is constant, and no disassociation occurs during preheating. Upper limits also increase linearly with temperature but only if the combustion is normal. If a rise in temperature tends to cause cool flames, the flammable range cannot be predicted reliably. With an increase in temperature the low composition area of normal flames disappears while the cool flame region is extending the envelope beyond the normal range [49].

2.1.4 Chemistry Dependence

Since a rigorous theory based on fundamental principles is not established, a lot of empirical rules or correlations were proposed for estimation purpose. Among them, the most famous one is Jone's rule or Lloyd's rule.

Jones [9] first observed that at any specified temperature, the ratio of the lower limit to the amount of combustible needed for stoichiometric reaction, is approximately constant. He further proposed several constants corresponding to families of fuels respectively (see Table 2.1).

Jones' rule assumes ratios of the lower limits of the individual constituents to the amount of oxygen required for theoretical perfect combustion are about the same, however, this is true only within a family of fuels. Under a similar reasoning, Shimy [62] used the number of carbon atoms in the molecule as the variable to correlate flammability limits within a family of fuels with limited success. If the fuel mixture came from a same family, the limits of mixtures of the constituents may be determined rather accurately by Jones' rule.

Lloyd [63] generalized Jones' rule into one universal constant, which is widely cited as "**Lloyd's Rule**"

$$x_L = 0.55 \cdot x_{st} = \frac{0.55}{1 + 4.773 \cdot C_O} \tag{2.8}$$

$$x_U = 3.5 \cdot x_{st} \tag{2.9}$$

Table 2.1 Typical values for x_L/x_{st} in Jones' rule

Methane	0.52	Methyl chloride	0.67
Ethane	0.57	Ethyl chloride	0.61
Propane	0.59	Propyl chloride	0.61

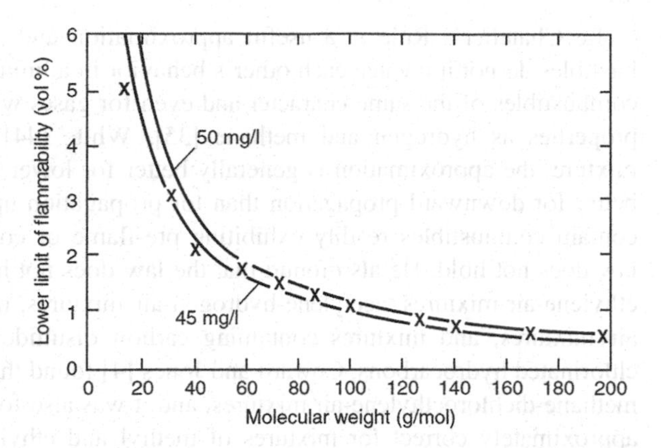

Fig. 2.2 Effect of molecular weight on lower limits of flammability of alkanes at 25 °C [64]

At room temperature and atmospheric or reduced pressure, the lower limits of flammability for most of the paraffin hydrocarbon series fall in the range from 45 to 50 mg combustible vapor per liter of air at standard conditions [43], which is corresponding to a limiting oxygen concentration of 0.11 in the mixture [17]. That means 45–50 mg of combustible vapor is sufficient to raise the temperature of one liter mixture from ambient to the critical flame temperature, which is another empirical rule used in industry ([64], see Fig. 2.2).

Other researchers also tried to establish the empirical relationship between LFL and UFL. White [44] found that the corresponding upper limit is roughly 3.5 times the lower limit. Spakowski [65] found that the upper limits were 7.1 times the lower limit. Zabetakis [43] proposes a square-root relationship, which has be used in the experimental work of Kondo [61].

$$x_U = 6.5\sqrt{x_L} = 4.8\sqrt{x_{st}} \tag{2.10}$$

2.1.5 Fuel Dependence

When multiple fuels are involved, Le Chatelier's Rule is the only tool for predicting mixture flammability. Originally, Le Chatelier [36] stated, "if we have, say, separate limit combustible air mixtures and mix them, then this mixture will also be a limit mixture", or a mixture of limit fuels is still a limit mixture. This statement was expressed as

$$\frac{x_1}{x_{L,1}} + \frac{x_2}{x_{L,2}} = 1 \tag{2.11}$$

where x_1, x_2 are the fuel concentrations in a mixture; $x_{L,1}, x_{L,2}$ are their individual flammable limits in air. It was Coward et al. [37] who extended Le Chatelier's rule into its present form (see Eqs. 4.22 and 4.23).

Le Chatelier's Rule is a useful approximation and implies that several combustibles do not influence each other's behavior in a limit mixture. It holds well for combustibles of the same character and even for gases with such different physical properties as hydrogen and methane [35]. White [44] found that for a binary mixture, the approximation is generally better for lower than for upper limits, and better for downward propagation than for propagation upwards. For mixtures that contain combustibles readily exhibiting pre-flame or cool flame combustion, the law does not hold. He also found that the law does not hold strictly for hydrogen-ethylene-air mixtures, acetylene-hydrogen-air mixtures, hydrogen sulfide-methane-air mixtures, and mixtures containing carbon disulfide. Also, in tests on some chlorinated hydrocarbons, Coward and Jones [4] found that the law did not hold for methane-dichloroethylene-air mixtures, and it was also found that the law was only approximately correct for mixtures of methyl and ethyl chlorides. It is therefore apparent that this mixture law cannot be applied indiscriminately, but must first be proved to hold for the gases being investigated [9].

Mashuga et al. [66] identified the assumptions in deriving Le Chatelier's rule,

1. the product heat capacities are constant;
2. the number of moles of gas is constant;
3. the combustion kinetics of the pure species are independent and unchanged by the presence of other combustible species;
4. the adiabatic temperature rise at the flammability limit is the same for all species.

Here assumption No. 3 and 4 are more important than the first two. No. 4 allows all species to compare with each other, while No. 3 allows them to be additive in terms of energy conservation. The major concern is whether fuels are synergistic or antagonistic to each other. Hydrogen is a typical example. In theory, due to its lower flame temperature and high diffusivity on flame structure, hydrogen will have a synergistic effect on other fuels. In reality, hydrogen is a common fuel and there is no specific limit in applying Le Chatelier's Rule for estimation purpose. The No. 1 assumption on constant heat capacities of combustion products can be dropped if scaled by a universal species [15].

White [44] further proposed the assumption that the ignition temperature of a gas mixture at concentrations similar to those present in a lower-limit mixture does not vary much with the concentration of the flammable gas. This temperature is constant no matter what is the initial fuel temperature, which pave the way for later theoretical work on temperature dependence. This assumption is widely found in most theoretical work, even implicitly used in deriving Le Chatelier's rule [66].

Generally, there are two types of methods to use Le Chatelier's rule covering diluents in a mixture. The popular method is grouping a fuel with a diluent into a pseudo fuel, and then Le Chatelier's Rule can be applied to pseudo-fuels only. So the diluted flammability diagram was used to compensate the inability of Le Chatelier's mixture law in dealing with diluents [67, 68]. A case study is provided by Heffington [69] on the flammability of CO_2-diluted fuels. Another estimation method is to determine the equivalency between gases for a mixture. Fuels are scaled by methane, inert gases are scaled by Nitrogen, and oxidizing gases are scaled by oxygen. By summing up each fuel/diluent/oxygen terms, the fuel and oxidizing potentials are determined and compared to determine the flammable state of a mixture [70, 71].

2.1.6 Pressure Dependence

An increase in pressure has little effect on the lower limit, though the upper limit may be widened; the extent of the reaction at the flame front is affected [5]. Based on the experimental data on natural gas (85–95 % methane and 15–5 % ethane) from Jones et al. [72], Zabetakis [43] suggested that the limits vary linearly with the logarithm of the initial pressure. That is

$$\begin{aligned} x_L &= 4.9 - 0.71 \log P(\text{atm}) \\ x_U &= 14.1 + 20.4 \log P(\text{atm}) \end{aligned} \tag{2.12}$$

With a standard error of estimate of 0.53 vol pct for x_L and 1.51 vol pct for x_U.

Figure 2.3 shows the flammability change according to high pressure. Lower flammability limits change little, since lower limits are controlled mainly by (the heat release of) fuel. Upper limits are changed significantly, which is related to the flame structure and combustion kinetics. In this specific area, experimental data are still limited.

For lower-than-ambient pressure, the flammable range is little changed until a certain lower pressure threshold is achieved. Then the flammable zone shrink to nil, consistent with our intuition that vacuum will not support ignition or flame propagation. Lewis and Von Elbe [32] produced a pressure-modified flammable zone in Fig. 2.4.

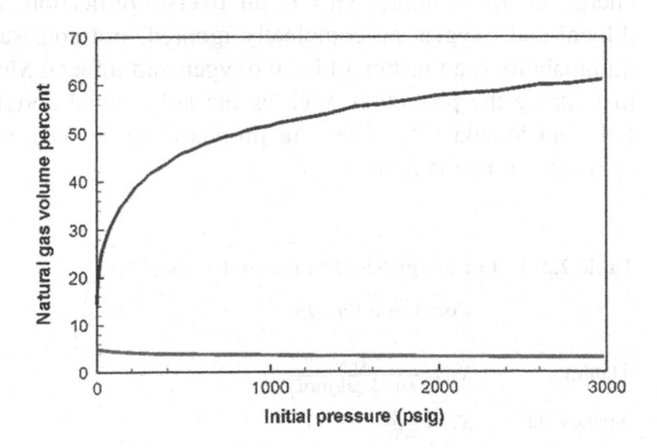

Fig. 2.3 Variation of flammability limits due to high pressure

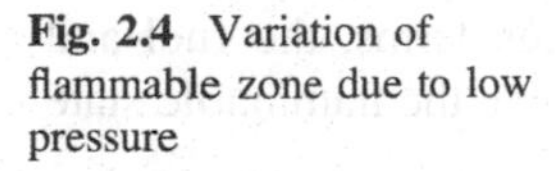

Fig. 2.4 Variation of flammable zone due to low pressure

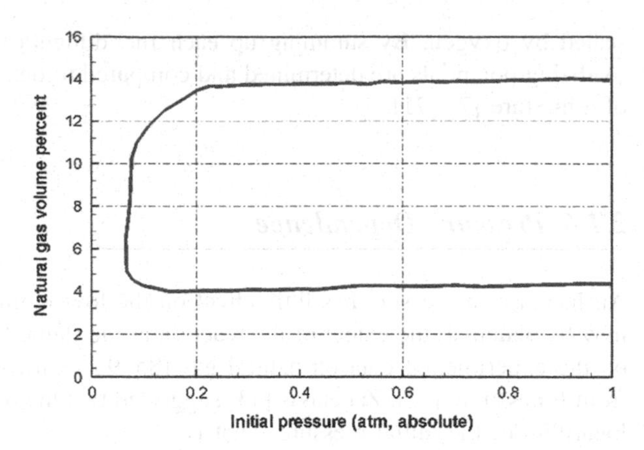

2.2 Correlations for Flammability

Based on the above principles, several correlations were proposed for estimating the flammability of a fuel or a mixture. They are falling into two categories, fuel-based or oxygen-based. The former is based on the heat of combustion (of the fuel), while the latter is based on the stoichiometric oxygen number, or oxygen calorimetry. Heat of combustion is a property related to the combustion process, or related to the ignitability of the fuel. Oxygen calorimetry is a property related to oxygen, or a property of the background air, or the explosibility. This fact itself shows the energy background for flammability is rather ambiguous and complex.

Let us check the fuel-based correlations in Table 2.2. From the modified Burgess-Wheeler's law, Hanley [47] proposed $X_L = \frac{11.2}{\Delta H_C(\text{kcal/mol})}$ for LFL, while $X_L = \frac{43.54}{\Delta H_C}$ is proposed by Spakowski [73]. These are slight variations of Burgess-Wheeler's law, which implicitly assumed that any fuel releases same amount of energy at lower limits. This is an oversimplification, since the information on diluent and oxygen are completely ignored, not consistent with the fact that the flammability is a function of local oxygen and diluent. More refinement is proposed to improve the prediction, such as the polynomial correlation proposed by Shieh [74] and Suzuki [75]. They are proposed based on a small set of data, so their applicable range is limited.

Table 2.2 List of energy-based LFL correlations [17]

	Correlation for LFL	Average error in estimation
Hanley	$X_L = \frac{11.2}{\Delta H_C(\text{kcal/mol})}$	0.119
Spakowski	$X_L = \frac{43.54}{\Delta H_C}$	0.069
Shieh	$x_L = 1145 \cdot \Delta H_c^{-0.8} - 0.38$	0.076
Suzuki	$x_L = -3.4 \cdot \Delta H_C^{-1} + 0.569 \cdot \Delta H_C + 0.0538 \cdot \Delta H_C^2 + 1.8$	n/a

Table 2.3 List of oxygen-based LFL correlations [17]

	Original correlation	Average error in estimation
Jones' rule/Lloyd's rule/Zabetakis [43]	$X_L = 0.55 \cdot x_{st} = \frac{0.55}{1+4.773 \cdot C_O}$	0.108
Britton [48]	$X_L = \frac{10}{C_O}$	0.070
Pintar [76]	$X_L = 0.512 \cdot x_{st} = \frac{0.512}{1+4.773 \cdot C_O}$	0.085
Hilado [81] (CHO compounds only)	$X_L = 0.537 \cdot x_{st} = \frac{0.537}{1+4.773 \cdot C_O}$	0.125
Monakhov [78]	$X_L = \frac{1}{4.679+8.684 \cdot C_O}$	0.077
Beyler [64] ($\phi = 0.5$)	$X_L = \frac{1}{1+9.546 \cdot C_O}$	0.094
Donaldson et al. [80]	$X_L = \frac{1}{5.2114+8.2069 \cdot C_O}$	0.091

Since flammability is the ability of background air to support flame propagation, a more reasonable choice for correlation is the stoichiometric oxygen number, or oxygen/fuel stoichiometric molar ratio. A list of correlations on this parameters is provided in Table 2.3. From Jones' rule, $X_L = 0.55 \cdot x_{st} = \frac{0.55}{1+4.773 \cdot C_O}$ is listed as an approximation of lower flammability limits. Britton [48] proposed a simpler form as $X_L = \frac{10}{C_O}$. Pintar [76] and Hilado [77] refined the correlation coefficients for a specific family of hydrocarbons. Monakhov [78], Beyler [79] and Donaldson et al. [80] realized that $x_L \cdot C_O$ is not a constant, so a constant is reserved in the denominator. This is consistent with the modified "k-constant" rule ($x_L \cdot \Delta H_C + \Delta H = K$). As we will see later in Sect. 4.4, $x_L \cdot \Delta H_C = K$ applies for ignitability, while $\lambda_2 \cdot \frac{\Delta H_C}{C_O} = K$ applies for explosibility. The difference is that the inerting contribution from the fuel is ignored in the former while the oxygen-based energy release is more dominant at critical limits.

For upper limits, no energy-based correlations are proposed, consistent with the fact that the upper limits are oxygen-limited. The fuel mainly plays the role as a diluent, while the availability of oxygen and oxygen calorimetry dominate the flame propagation process. Due to the scattering of experimental data, all correlations are rough at predicting upper limits. Details of their performance can be found in Table 2.4, with a prediction error identified in [17].

2.3 ISO10156 Method for Predicting Mixture Flammability

Currently, the professional method to deal with multiple fuels and diluents is provided in ISO10156 [70], which contains the testing method and calculation methods for mixture flammability. The calculation method uses flammability limit data and so-called Tci values of flammable gases and vapors for inputs. These data

Table 2.4 List of oxygen-based UFL correlations [17]

	Original correlation	Average error in estimation
Zabetakis [43]	$x_U = 3.3 \cdot x_{st} = \frac{3.3}{1+4.773 \cdot C_O}$	1.227
Pintar [76]	$x_U = 3.8 \cdot x_{st} = \frac{3.8}{1+4.773 \cdot C_O}$	2.345
Monakhov [78]	$\begin{cases} x_U = \frac{1}{0.56+1.55 \cdot C_O}, & \text{for } C_O \leq 7.5 \\ x_U = \frac{1}{6.554+0.768 \cdot C_O}, & \text{for } C_O > 7.5 \end{cases}$	0.289
Donaldson et al. [80]	$x_U = \frac{1}{1.2773+1.213 \cdot C_O}$	0.303
Beyler [64] ($\phi = 3.0$)	$x_U = \frac{1}{1.0+1.591 \cdot C_O}$	0.182

are mostly taken from the database CHEMSAFE that contains recommended safety characteristics for flammable gases, liquids and dusts.

The Tci values are taken under atmospheric conditions with air as an oxidizer, which is also called the maximum permissible flammable gas concentration (MXC). MXC is the largest fraction of the flammable gas for which this gas mixture cannot be ignited, irrespective of the amount of flammable gas being added [70]. As we will see later (in Chap. 5), MXC is a concept of ignitability, equivalent to LFC or OSFC in the literature. In contrast, the coefficients of equivalency relative to nitrogen (K value) must be calculated using several flammable-inert gas system. The K value for the mixture is calculated as

$$K_{i,j} = \frac{MXC_{i,j} \cdot \left(\frac{100}{Tc_i} - 1\right)}{\left(100 - MXC_{i,j}\right)} \tag{2.13}$$

where i is the index of the flammable gas and j is the index of the inert gas, different from Nitrogen which is used in the flammable/inert/air mixture. Table 2.5 lists coefficient of equivalency (Ki) for some common diluents, which are close to the concept of Quenching Potential to be introduced in next chapter. Table 2.6 lists Tci values for common flammable gases, which is another name for LFC or OSFC for nitrogen inertion, to be introduced and discussed in Chap. 5.

Table 2.5 Coefficient of equivalency (Ki), or nitrogen-equivalency

Gas	N_2	CO_2	He	Ar	Ne	Kr	Xe	SO_2	SF_6	CF_4
K_i	1	1.5	0.5	0.5	0.5	0.5	0.5	1.5	1.5	1.5

Table 2.6 Tc_i for flammable gases, or the limiting fuel concentration for nitrogen-inerting

Gas	H2	CO	Methane	Ethane	Butanes	Ethylene	Propane	Propene's	Acetylene	n-Hexane
Tc_i	5.7	20	8.7	7.6	5.7	6	6	6.5	4	3.5

The mixture has to be classified as FLAMMABLE if

$$\sum_{i=1}^{n} A_i\left(\frac{100}{Tc_i}-1\right) \leq \sum_{k=1}^{p} B_k K_k \tag{2.14}$$

where

A_i Mole fraction of the flammable component I in the mixture in mol%
n Number of flammable components
B_k Mole fraction of the inert component k in the mixture in mol%
p Number of inert components
K_k Nitrogen equivalence coefficient of the inert component
Tc_i Threshold for flammability of the flammable component I in the mixture with nitrogen

The physical meaning of Eq. 2.14 is the fuel-needed diluent (nitrogen, left-hand side) should be less than existing diluent (nitrogen, right-hand side) to keep the mixture non-flammable.

A variation of Eq. 2.13 is supplied [70] as

$$\frac{1}{MXC_{mixture}} = \frac{\left(\frac{100}{Tc_i}-1\right)}{B_{K1} \times K_{K1} + B_{K2} \times K_{K2}} + 1 \tag{2.15}$$

Which can be used to find the critical fuel concentrations in response to multiple diluents.

2.4 Flammability Diagrams

The flammability problem of a mixture is difficult, since it typically involves three components with dual functions in a combustion reaction. Fuel is not only a source of energy, but also a heat absorber during the ignition process. Oxygen is not only a source of energy (together with fuel), but also a heat absorber affecting the flame temperature. Only nitrogen is a typical diluent, without any heating role involved. Because of such a ternary combustion system, the concept of flammability is difficult to present without the help of a flammability diagram. However, various diagrams are proposed to demonstrate the inerting and diluting process, all with some limitations over the past century. Depending on its application field and purpose, there are four diagrams are typically used in industry.

2.4.1 Standard Flammability Diagram

The first diagram is called standard flammability diagram, which uses the combination of diluent/fuel to describe the state of a mixture. Since the diluent concentration is serving as a major input variable, it is commonly used to compare the agent effectiveness, in chapters of suppression theory [82]. It is also used in flammability theory, to demonstrate the role of a diluent on changing flammable envelopes [64]. For practical safe handling of gases, since the concentration measurement is not convenient as the volume measurement (nitrogen concentration is derived from oxygen measurement, not directly measured), the diluted diagram is used instead of the standard diagram, which can be converted to each other easily. The critical nose point in a standard flammability diagram is the inertion point, which tells the Minimal Inerting Concentration (MIC) (shown in Fig. 2.5). Most of flammability data in Zabetakis' BOM Bulletin [43] are presented in this form, while it provides less information as compared to other diagrams.

2.4.2 Diluted Flammability Diagram

Using the diluent/fuel ratio and the diluent/fuel fraction as inputs, we have a diluted flammability diagram. Its original purpose is to pair a fuel and a diluent into a pseudo fuel, then Le Chatelier's rule can be applied to a mixture of pseudo fuels [83, 84]. As Le Chatelier's rule is the only hand-calculation tool in industry, and the industry prefers the volume measurement in dilution, diluted flammability diagram gains a special position on safe handling of flammable gases.

From the flammable envelope, two critical lines can be derived, MMR (Maximum Molar (diluent/fuel) Ratio) and MMF (Maximum allowable fuel Mixture Fraction), which provides a rectangular boundary encompassing the flammable envelope. Note, MMR and MMF may not be taken from a same point on

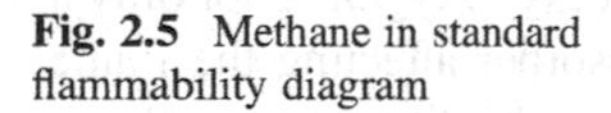
Fig. 2.5 Methane in standard flammability diagram

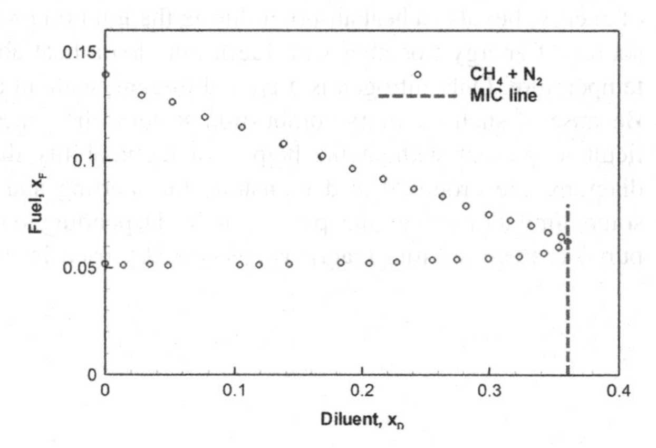

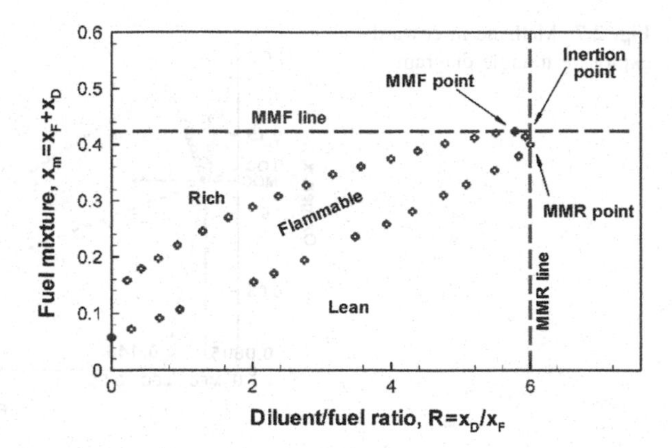

Fig. 2.6 Methane in diluted flammability diagram

the envelope. Theoretically, there is only one inertion point, while in experiments, there is a maximum MMF (inertion) point and a maximum MMR (dilution) point, on MMF and MMR lines respectively (as shown in Fig. 2.6). In this perspective, the inertion point in an experimental flammability diagram is fictitious.

However, this fictitious inertion point provides valuable information (MMR and MMF) on explosibility and ignitability for other critical points, so this diagram is fundamental to any flammability related operations.

2.4.3 Explosive Triangle Diagram

When multiple diluents are included in a mixture, such as gases from a mine fire, the flammable state of such a mixture is difficult to present in any diagrams with a definite axis on diluent. Instead, Coward explosive triangle was proposed to use the oxygen level in the mixture, avoiding the complexity induced by multiple diluents [68]. If drawing a line from the 100 % fuel point, tangent to the experimental flammable envelope, this is called LOC line (Limiting Oxygen Concentration) with a cross point on Oxygen axis as LOC point. Similarly drawing a straight line from the normal air (20.95 % O_2 + 79.05 % N_2) point tangent to the flammable envelope, this is called LFC (Liming Fuel Concentration) line. The cross point on fuel axis is called LFC point. In practice, the iso-oxygen line tangent to the envelope is called MOC (Minimum allowable Oxygen Concentration to support flame propagation), while the fuel concentration at the tangent point of LFC line on flammable envelope is called Minimum Fuel Concentration (MFC).

If the background air is inerted by a diluent to LOC point, the evaporation/addition of fuel will move this point from LOC to MOC along the LOC line. If a fuel stream is already diluted by a diluent to its LFC point, then mixing with air will decrease this fuel concentration further from LFC to MFC along the LFC line. In theory, LFC and LOC are better and fundamental critical targets of inertion.

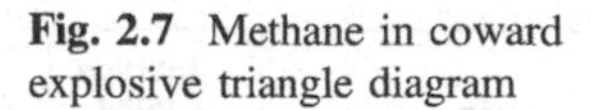

Fig. 2.7 Methane in coward explosive triangle diagram

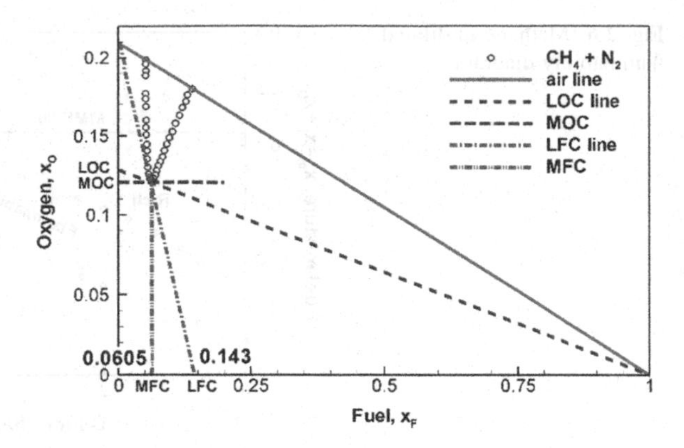

In practice, MOC and LFC are recommended for safe-operation (dilution or purge) targets (as shown in Fig. 2.7).

When dealing with a complex mixture, such as the gases from a burning mine fire, various fuels are lumped into a fuel equivalent (usually methane), while contributions of diluents will be lumped into an oxygen limit, plotted in the diagram as the nose point. Various schemes are proposed to perform the conversion more effectively [85]. However, the fundamental theory behind such a conversion is provided in Chap. 7.

2.4.4 Ternary Flammability Diagram

The major problem of a diluted flammability diagram is that it is difficult to demonstrate the dilution and purge processes (not a straight line), as composite parameters are used. A ternary flammability diagram was proposed [43] and gained a popular position in industry for guiding dilution and purge operations. The advantage of a ternary diagram is that all data are directly readable and oxygen-enriched atmosphere is allowed, while the disadvantages is that a conversion scheme is needed to plot the data. Crowl [86] listed some engineering rules to use this diagram better. Another disadvantage of this diagram is that oxygen only occupies 1/5 of air, so a large part of domain is useless if we are dealing with fuels burning in air. How to make use of space more effectively is a skill requiring special training.

Starting with the 100 % fuel point, draw a line tangent to the flammable envelope, it is called the LOC line. If an iso-oxygen line is tangent to the flammable envelope, this is called the MOC line. Starting with the air point (21 % O_2 and 79 % N_2), draw a line tangent to the flammable envelope, which is called the LFC line. If an iso-fuel line is tangent to the flammable envelope, this is called the MFC line. These four lines are demonstrated in Fig. 2.8. Again, MFC point and MOC point may not be a same point, though both are representing the inertion point.

2.5 Problems and Solutions

2.5.1 ISO10156 Method

Problem 2.1 Tc_i method

A gaseous mixture of 7 % of hydrogen in CO_2, is it flammable in air? [71]

Solution:

K_k, Nitrogen equivalency of CO_2, is 1.5.

Therefore 93 % CO_2 is equivalent to 93 × 1.5 or 139.5 % nitrogen.

Normalized hydrogen concentration is $x_{H_2} = \frac{7}{7+139.5} = 4.78\,\%$. The Tc_i value of Hydrogen is 5.7 > 4.78, therefore the mixture is not flammable.

Problem 2.2 Tc_i method for multiple fules.

A mixture has methane/propane (20:80 mol%) altogether 5.6 %. The rest is nitrogen. Is it flammable in air?

Solution:

Tc_i values of methane and propane are retrieved from Table 2.6 as 8.7 and 3.7 respectively. K_k for nitrogen is 1 by definition. B_k for nitrogen is given as 1–5.6 % = 94.4 %. $A_{methane} = 20\,\% \times 5.6\,\% = 1.12\,\text{mol}\%$, $A_{propane} = 80\,\% \times 5.6\,\% = 4.48\,\text{mol}\%$. So we have

$$\sum_{i=1}^{n} A_i\left(\frac{100}{Tc_i} - 1\right) = \left[1.12 \times \left(\frac{100}{8.7} - 1\right) + 4.48 \times \left(\frac{100}{3.7} - 1\right)\right] = 128.35\,\% \leq \sum_{k=1}^{p} B_k K_k = 1 \times 94.4\,\%$$

Since the inequality is not fulfilled, the mixture is not fully diluted, or is classified as FLAMMABLE.

Problem 2.3 Tc_i method for multiple diluents

A mixture has a methane fraction of 16.4 %. The rest is carbon dioxide/nitrogen (70:30). Is it flammable in air?

Solution:

Tc_i value of methane is retrieved from Table 2.6 as 8.7.

The limiting fuel fraction, MXC is computed from

$$\frac{1}{MXC_{mixture}} = \frac{\left(\frac{100}{Tc_i} - 1\right)}{B_{K1} \times K_{K1} + B_{K2} \times K_{K2}} + 1 = \frac{\left(\frac{100}{8.7} - 1\right)}{0.7 \times 1.5 + 0.3 \times 1} + 1 = 8.77$$

So the limiting fuel fraction is

$$MXC = \frac{1}{8.77} = 11.4\,\% < 16.4\,\%$$

So this mixture is non-flammable.

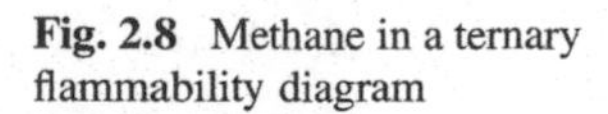

Fig. 2.8 Methane in a ternary flammability diagram

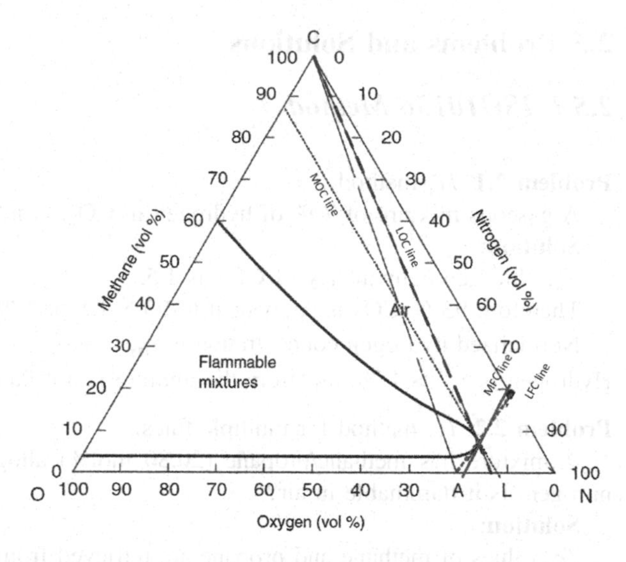

Problem 2.4 Tc_i method for multiple fuels and multiple diluents

A flammable mixture has 2 % Hydrogen, 8 % methane, 65 % Helium, and 25 % Argon, is it flammable in air? [71]

Solution:

Using nitrogen-equivalent coefficients K_k, this mixture is equivalent to 2 % H_2 + 8 % CH_4 + 0.55 × 25 % Ar + 0.9 × 65 % He = 2 % H_2 + 8 % CH_4 + 13.8 % N_2 + 58.5 % N_2 = 82.3 %

Rescale the mixture by 100/82.3 = 1.215, we have the mixture as 2.43 % H_2 + 9.73 % CH_4 + (58.5 + 13.8) × 1.213 % N_2

Here Tc_i for hydrogen and methane is 5.7 and 8.7 respectively.

$$\sum_{i=1}^{n} A_i\left(\frac{100}{Tc_i}-1\right) = \left[2\times\left(\frac{100}{5.5}-1\right)+8\times\left(\frac{100}{8.7}-1\right)\right] = 118.3\,\% \le \sum_{k=1}^{p} B_k K_k = 0.55\times 25\,\% + 0.9\times 65\,\% = 75.25$$

Since the above inequality does not hold true, the criterion for a non-flammable gas mixture is not fulfilled and this particular gas mixture is considered flammable.

2.5.2 Operations in a Ternary Diagram

Problem 2.3 Use 5 points (MOC = 12 %, LFL(O_2) = 5 %, UFL(O_2) = 61 %, LFL (air) = 5 %, UFL(air) = 15 %) to reconstruct the flammable envelope of methane (Fig. 2.9).

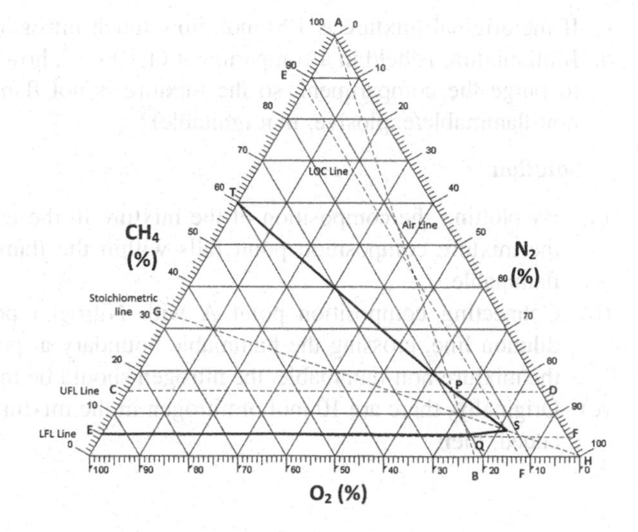

Fig. 2.9 5-pts to determine the flammable zone for methane

Solution:

Step 1. Plot UFL(O_2) = 60 % and LFL(O_2) = 5 % in fuel axis as Point T/E respectively.

Step 2. Use nitrogen concentration of 79 % in air (point B), draw an airline AB.

Step 3. Use UFL(air) = 15 % to draw an UFL line CD, crossing the airline AB with point P.

Step 4. Use LFL(air) = 5 % to draw a LFL line EF, crossing the airline AB with point Q.

Step 5. Find the stoichiometric value through the balanced equation, $CH_4 + 2O_2 = CO_2 + H_2O$, so the stoichiometric fraction of methane in methane/oxygen mixture is 1/(1 + 2) = 0.33. Connect point G (33 % CH_4) with H, is the stoichiometric line GH.

Step 6. Use MOC = 12 % to draw an LOC line EF, crossing the stoichiometric line GH with point S.

Now in Fig. 2.9, the triangle PQS is the flammable envelope in air, while the polygon TPSQE is the flammable envelope of methane.

2.5.3 *Reading a Ternary Diagram*

Problem 2.5 There is a methane mixture composed of 50 % methane, 40 % O_2 and 10 % Nitrogen.

a. Is this mixture flammable?
b. If this mixture is flammable, how to make it non-flammable by adding nitrogen for inertion?

c. If the original mixture is 100 mol, how much nitrogen is needed?
d. If all mixture is held in a compartment of 100 m^3, how much nitrogen is needed to purge the compartment, so the mixture is not flammable (out-of-envelope, non-flammable/explosive, non-ignitable)?

Solution:

(a) By plotting the composition of the mixture in the ternary diagram (point A), the mixture composition point falls within the flammable envelope, so it is flammable.
(b) Connecting composition point A with Nitrogen point, we have a straight dilution line, crossing the flammable boundary at point B. So in order to get the mixture non-flammable, the nitrogen should be increased from 10 to 30 %.
(c) Originally, there are 10 mol of nitrogen in the mixture, assume y mole is to be added, then

$$30\,\% = \frac{10 + y}{100 + y} \rightarrow \mathrm{y} = 28.6\ \mathrm{mol}$$

That means the system will be non-flammable if added with 28.6 mol of nitrogen.

(d) From the diagram (Fig. 2.10), the out-of-flammable zone state is reached if the fuel concentration is dropped from 50 to 40 % (point B), so the purge requirement is $V_{N_2} = -V_0 \cdot \ln\left(\frac{x_F}{x_{F,0}}\right) = -100 \times \ln\left(\frac{0.4}{0.5}\right) = 22.3\ \mathrm{m}^3$. That means 22.3 m^3 of nitrogen will be introduced into the compartment, so the fuel concentration will be dropped to 40 %.
(e) From the diagram (Fig. 2.10), the non-flammable/explosive zone state is reached if the fuel concentration is dropped from 50 to 18 % (point C), so the purge requirement is $V_{N_2} = -V_0 \cdot \ln\left(\frac{x_F}{x_{F,0}}\right) = -100 \times \ln\left(\frac{0.18}{0.5}\right) = 102.2\ \mathrm{m}^3$. That means 102.2 m^3 of nitrogen will be introduced into the compartment, so the fuel concentration will be dropped to 18 %.
(f) From the diagram (Fig. 2.10), the non-ignitable zone state is reached if the fuel concentration is dropped from 50 to 10 % (point D), so the purge requirement is $V_{N_2} = -V_0 \cdot \ln\left(\frac{x_F}{x_{F,0}}\right) = -100 \times \ln\left(\frac{0.1}{0.5}\right) = 160.9\ \mathrm{m}^3$. That means 160.9 m^3 of nitrogen will be introduced into the compartment, so the fuel concentration will be dropped to 10 %.

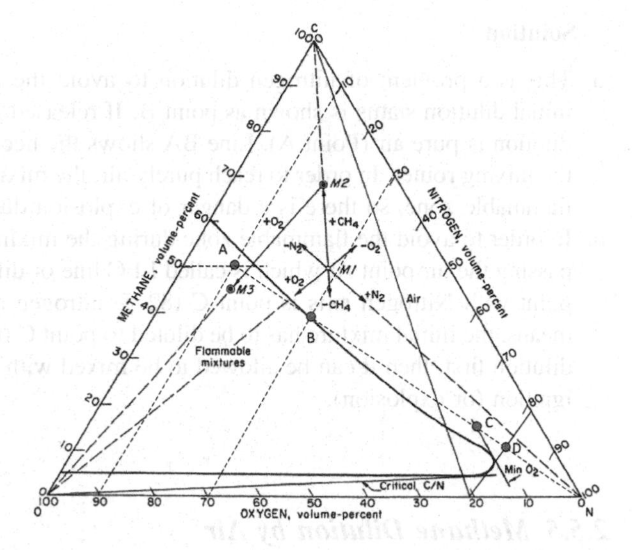

Fig. 2.10 Dilution by nitrogen in methane's flammability diagram

2.5.4 Safe Dilution of a Flammable Gas Mixture

Problem 2.6 A vessel contains a gas mixtures composed of 50 % methane and 50 % nitrogen. If the mixture escapes from the vessel and mixes with air, will it become flammable? How to make it strictly non-flammable during the dilution process?

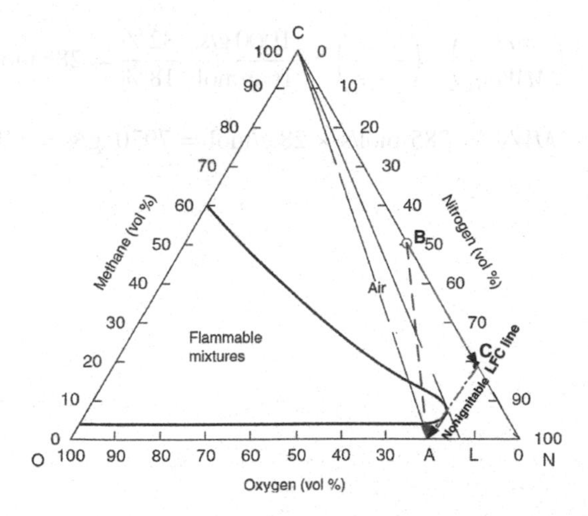

Solution:

a. This is a problem of nitrogen dilution to avoid the flammable envelope. The initial dilution status is shown as point B. If released into air, the final status of dilution is pure air (Point A). Line BA shows the necessary mixing process (or the mixing route). In order to reach purely air, the mixing line BA is crossing the flammable zone, so there is a danger of explosion during the mixing process.
b. In order to avoid the flammable zone during the mixing, a tangent line is drawn passing the air point A, which is called LFC line or dilution line later. The cross point with Nitrogen axis is point C (82 % nitrogen and 18 % methane). That means, the initial mixture has to be diluted to point C (non-ignitable) by nitrogen dilution first, then it can be allowed to be mixed with air without any danger of ignition (or explosion).

2.5.5 *Methane Dilution by Air*

Problem 2.7 A 1 kg/s flow of methane is being dumped into the atmosphere. How much nitrogen must be mixed with methane to avoid a flammable mixture in the open?

Solution:

From the previous ternary diagram, the inertion is realized through diluting the mixture from 100 % methane to 18 % methane with addition of 82 % of Nitrogen. So the nitrogen molar flow rate is

$$\dot{n}_{N_2} = \left(\frac{\dot{m}_{CH_4}}{MW_{CH_4}}\right) \cdot \left(\frac{C_{N_2}}{C_{CH_4}}\right) = \frac{1000\,\text{g/s}}{16\,\text{g/mol}} \cdot \frac{82\,\%}{18\,\%} = 285\,\text{mol/s}$$

$$\dot{m}_{N_2} = \dot{n}_{N_2} \cdot MW_{N_2} = 285\ \text{mol/s} \times 28\,\text{g/mol} = 7970\ \text{g/s} = 7.97\ \text{kg/s}$$

Chapter 3
Combustion Fundamentals

3.1 Thermochemistry

3.1.1 Chemical Reactions and Stoichiometry

Combustion is a sequence of exothermic chemical reactions between a fuel and an oxidizer accompanied by the release of heat and conversion of chemical species. Since combustion is a typical chemical reaction, all chemical rules apply in a combusting reaction. Let's start with a stoichiometric reaction to introduce the principles in thermochemistry.

An estimation starts with a chemically-balanced equation for the conservation of atoms. It commonly expressed in this generic form.

$$C_aH_bO_cN_d + \nu_{air}(O_2 + 3.76N_2) \rightarrow \nu_{CO_2}CO_2 + \nu_{H_2O}H_2O + \nu_N N_2 \tag{3.1}$$

For the case of an ideal combustion, or a complete reaction, we call it a stoichiometric reaction. **Stoichiometry** is a branch of chemistry that deals with the relative quantities of reactants and products in chemical reactions. In **a stoichiometric chemical reaction**, the oxygen supply and the fuel supply are proportional to a fixed number, commonly called **stoichiometric oxygen number**. The reaction is balanced if the stoichiometric oxygen number is reached, and the products have similar stoichiometric coefficients based on the conservation of atoms. To balance the generic Eq. 3.1, we have

$$\begin{aligned} \nu_{air} &= a + \frac{b}{4} - \frac{c}{2} \\ \nu_{CO_2} &= a \\ \nu_{H_2O} &= \frac{b}{2} \\ \nu_N &= 3.76\nu_O + \frac{d}{2} \end{aligned} \tag{3.2}$$

T. Ma, *Ignitability and Explosibility of Gases and Vapors*,
DOI 10.1007/978-1-4939-2665-7_3

Sometimes, the air/fuel ratio for a complete reaction on a mass basis, $r = \left(m_f/m_{O_2}\right)_{stoich}$, is used to represent the reaction stoichiometry. The molecular oxygen/fuel ratio, v_{air}, can be derived by $v_{air} = r \cdot \frac{M_f}{M_{air}}$.

3.1.2 Equivalence Ratio and Non-stoichiometric Equations

If not all fuel or oxidant are consumed in a reaction, a new variable is introduced as the real fuel/air ratio over the stoichiometric fuel/oxygen ratio, which is called **equivalence ratio**. This concept is used to characterize the deviation from a stoichiometric reaction.

$$\phi = \frac{(\text{fuel/air})_{\text{actual}}}{(\text{fuel/air})_{\text{stoich}}} = (\text{fuel/air})_{\text{actual}} \cdot 4.76 \cdot v_{air} \tag{3.3}$$

Equivalence ratio can also be defined as a mass ratio.

$$\phi = \frac{\left(m_f/m_{O_2}\right)_{actual}}{\left(m_f/m_{O_2}\right)_{stoich}} = \frac{\left(m_f/m_{O_2}\right)_{actual}}{r} \tag{3.4}$$

The purpose of introducing equivalence ratio is to characterize the deviation of a reaction from a stoichiometric reaction. For a **fuel-rich (oxygen-lean) reaction**, the residual fuel is still available in combustion products without any oxygen to react. This means that the real fuel supply is more than what is needed or the stoichiometric requirement, i.e. $\phi > 1$. Similarly for the **fuel-lean (oxygen-rich) reaction**, the fuel is fully consumed with oxygen level being non-zero, i.e. $\phi < 1$. Equivalence ratio is most useful in characterizing the conditions for smoke production.

For those non-stoichiometric reactions, we have to compute the coefficients based on the equivalence ratio. Here we will rewrite the reaction with a new set of coefficients, which are variables based on the stoichiometric coefficients.

$$\begin{aligned} &C_aH_bO_cN_d + \mu_{air}(O_2 + 3.773N_2) \\ &\rightarrow \mu_{CO_2}CO_2 + \mu_{H_2O}H_2O + \mu_N N_2 + \mu_O O_2 + \mu_f C_aH_bO_cN_d \end{aligned} \tag{3.5}$$

For a fuel-lean condition ($\phi < 1$), the full consumption of fuel is the limiting factor on energy release. Thus, the residual oxygen in products is a function of initial fuel concentration x_f, as shown in Eq. 3.6. For a fuel-rich condition ($\phi > 1$), the full consumption of oxygen is the limiting factor on energy release. So the residual fuel in products is a function of initial oxygen concentration x_O, as shown in Eq. 3.6.

$$\begin{array}{ll}
\phi < 1 & \phi > 1 \\
\mu_{air} = \frac{\nu_{air}}{\phi} = \frac{1-x_f}{4.773 \cdot x_f} & \mu_{air} = \frac{\nu_{air}}{\phi} = \frac{1-x_f}{4.773 \cdot x_f} \\
\mu_{CO_2} = a & \mu_{CO_2} = (1-\mu_f) \cdot a \\
\mu_{H_2O} = \frac{b}{2} & \mu_{H_2O} = (1-\mu_f) \cdot \frac{b}{2} \\
\mu_N = 3.773 \cdot \mu_{air} + \frac{d}{2} & \mu_N = 3.773 \cdot \mu_{air} + (1-\mu_f) \cdot \frac{d}{2} \\
\mu_O = \mu_{air} - \mu_{CO_2} - \frac{\mu_{H_2O}}{2} & \mu_O = 0 \\
\mu_f = 0 & \mu_f = 1 - \frac{\mu_{air}}{\nu_{air}}
\end{array} \tag{3.6}$$

Note 1/0.2095 = 4.773 and 1/0.21 = 4.76. Both constants are used in various literatures on flammability. Here 4.773 is used for improved resolution.

For stoichiometric reactions, two constants are generally used to represent the reaction and serve as the only variable for correlations. One is the stoichiometric oxygen/fuel ratio, C_O. The other is the heat of combustion, to be introduced next. This is sometimes called **the stoichiometric oxygen/fuel molar ratio**, or simply **stoichiometric oxygen number**. This number is representing the oxygen consumption in an ideal reaction, so it is commonly used for representing a chemical reaction or predicting the energy release.

$$C_O = \nu_{air} = a + \frac{b}{4} - \frac{c}{2} \tag{3.7}$$

Similarly, we can define a stoichiometric mixture/fuel ratio, C_{st}, which is a slight modification to C_O.

$$C_{st} = 1 + 4.773 \cdot C_O \tag{3.8}$$

Caution should be made on C_{st}, as some authors [87] used this symbol for the stoichiometric concentration of a fuel, which is defined as

$$x_{st} = \frac{1}{C_{st}} = \frac{1}{1 + 4.773 \cdot C_O} \tag{3.9}$$

3.1.3 Heat of Formation

Once the molecular ratio of fuel/oxygen is established, the next concept to be introduced is Heat of formation. Here, **standard heat of formation** of a compound is the change of enthalpy that accompanies the formation of one mole of a substance in its standard state from its constituent elements in their standard states (the most stable form of the element at 1 bar of pressure and the specified temperature, usually 298.15 K or 25 °C). Its symbol is ΔH_f.

For a typical combustion

$$C(graphite) + O_2(gas) \rightarrow CO_2(gas) \tag{3.10}$$

The reactants are naturally available elements, so their heat of formations are designated as zero, while $\Delta H_f^{298}(CO_2) = -393.5$ kJ/mol. Here the negative sign shows that CO_2 is a more stable chemical than its reactants. To get this product, the reaction is releasing energy. As a material property, heats of formation for some common species are listed in Table 3.1.

There are some notes on applying this concept. First, the heat of formation is based on the reference state. Secondly, heat of formation can be zero or negative given the chemical process to get this agent. Once the state (gas/liquid/solid) and temperature is fixed, each component has one fixed heat of formation as the characterizing feature. If the reactants are naturally available, such as nitrogen, its heat of formation is defined as zero. Thirdly, most fuels are formed by absorbing energy (as they release energy while decomposing), so they have a negative heat of formation.

Table 3.1 Heat of formation for common species ([88])

Substance	Formula	State	$\Delta\tilde{h}_f^\circ$ (kJ/mol)
Oxygen	O_2	g	0
Nitrogen	N_2	g	0
Graphite	C	s	0
Diamond	C	s	1.88
Carbon dioxide	CO_2	g	−393.5
Carbon monoxide	CO	g	−110.5
Hydrogen	H_2	g	0
Water	H_2O	g	−241.8
Water	H_2O	1	−285.9
Chlorine	Cl_2	g	0
Hydrogen chloride	HCl	g	−92.3
Hydrogen cyanide	HCN	g	+135.1
Methane	CH_4	g	−74.9
Propane	C_3H_8	g	−103.8
n-Butane	C_4H_{10}	g	−124.7
n-Heptane	C_7H_{16}	g	−187.8
Benzene	C_6H_6	g	+82.9
Formaldehyde	CH_2O	g	−115.9
Methanol	CH_4O	g	−201.2
Methanol	CH_4O	1	−238.6
Ethanol	C_2H_6O	1	−277.7
Ethylene	C_2H_4	g	52.5

3.1.4 Heat of Combustion

Heat of formation is a material property closely related to its chemical composition. In a chemical process, **Hess's law of constant heat summation** states that the change in enthalpy depends only on the initial and final states of the system and is independent of the reaction routes. Thus we can have a process quantity defined as the difference of energy summations between reactants and products, which is commonly called the **Heat of combustion**, defined as

$$\Delta H_C = \sum v_i \cdot \Delta H^0_{f,i}\Big|_{\text{reactants}} - \sum v_j \cdot \Delta H^0_{f,j}\Big|_{\text{products}} \tag{3.11}$$

The heat of combustion of solids or liquids is usually measured in a device known as an oxygen bomb calorimeter, in which a known mass of fuel is burnt completely in an atmosphere of pure oxygen, and the heat release is measured by means of the temperature rise to the surrounding water-bath. Table 3.2 lists some typical heat of combustion for common fuels fully burnt in air. Note, heat of combustion is a process property, different reaction route may lead to different heats of combustion, as shown in sample problems in Sect. 3.4.4.

3.1.5 Oxygen Calorimetry

Though the fuel is an energy-carrier in a chemical reaction, oxygen plays a vital role on releasing the energy from the fuel. Burgess and wheeler [6] first observed that the heat of combustion of one mole of fuel at the lean limit is nearly constant, for

Table 3.2 Heat of combustion for some common fuels fully burnt in air [88]

Fuel	ΔH_c (kJ/mol)	ΔH_c (kJ/g)	ΔH_c [kJ/g(O_2)]	ΔH_c [kJ/g(air)]
Carbon monoxide (CO)	283	10.10	17.69	4.10
Methane (CH_4)	800	50.00	12.54	2.91
Ethane (C_2H_6)	1423	47.45	11.21	2.96
Ethene (C_2H_4)	1411	50.53	14.74	3.42
Ethyne (C_2H_2)	1253	48.20	15.73	3.65
Propane (C_3H_8)	2044	46.45	12.80	2.97
n-Butane (n-C_4H_{10})	2650	45.69	12.80	2.97
n-Pentane (n-C_5H_{12})	3259	45.27	12.80	2.97
n-Octane (n-C_8H_{18})	5104	44.77	12.80	2.97
c-Hexane (c-C_6H_{12})	3680	43.81	12.80	2.97
Benzene (C_6H_6)	3120	40.00	13.06	3.03
Methanol (CH_3OH)	635	19.83	13.22	3.07
Ethanol (C_2H_5OH)	1232	26.78	12.88	2.99
Acetone (CH_3COCH_3)	1786	30.79	14.00	3.25

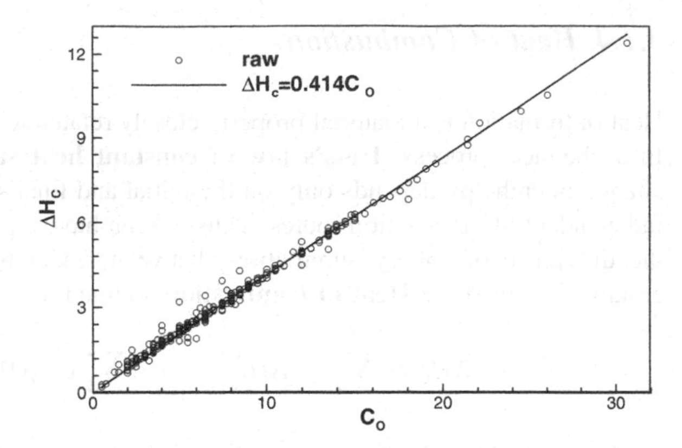

Fig. 3.1 Oxygen calorimetry for some common liquid fuels [17]

fuel air mixture at room temperature and pressure. Thornton [89] confirmed the constancy between the oxygen and the heat of combustion. Huggett [90] found this rule applies to most of solid fuels. Now called oxygen calorimetry, this rule has found many applications in theories and experiments.

From Fig. 3.1, we can see that heat of combustion is a linear function of stoichiometric oxygen coefficient (as shown in Eq. 3.12). The correlation is forced to have a zero intercept, since we want to get a single constant following the theory.

$$\Delta H_C = 0.414 \cdot C_O \tag{3.12}$$

Equation 3.12 means that one mole of oxygen will release 0.414 MJ of energy, which translates to 12.94 MJ/kg for oxygen calorimetry. This value is close to the commonly accepted value of 13.1 MJ/kg [90]. This means, one mole of oxygen in reaction will release 0.414 MJ of energy, no matter what hydrocarbon fuel is involved. So we can estimate the total energy by measuring the oxygen depletion, instead of measuring the mass loss of fuel directly. The fuel loss is difficult for direct measurement, since the fuel may not be a pure substance, and the combustion process may be incomplete and far from stoichiometry. By gauging the oxygen concentration in the flue gases (or combustion products) and recording the mass flow rate, the total oxygen consumption rate can be estimated. This is translated into a total energy release rate, which is the theoretical basis for most calorimetry devices.

For those chemicals with an oxygen calorimetry well above 13.1 MJ/kg, they are treated as explosive materials [59], which are not common fuels, so not covered here.

3.2 Adiabatic Flame Temperature

Flame temperature is important in understanding critical behaviors such as ignition and extinction. Each fire has its own characteristic flame temperature, as shown in Table 3.3. Adiabatic flame temperature by definition is the maximum possible

Table 3.3 Common flame temperatures in a fire

Flame scenarios	Temperature
Theoretical extreme (irreversible, stoichiometric, no diluent)	~6000 K
Highest flame temperature (reversible)	~3000 K
Burning in stoichiometric in air	~2400 K
Inertion at extinction	~1800 K
Ignition at LFL	~1600 K
Fire temperature	900 C–1000 °C
Large pool fire temperature	1100–1200 °C
Flame tip	320–550 °C
Warehouse storage rack fire	870 °C
Room fire, ceiling layer	>600 °C
Room fire, peak value	1200 °C
Post-flashover fire	900–1000 °C

temperature achieved by the reaction in a constant pressure process [91]. The classical method to compute the adiabatic flame temperature is proposed by Gordon and McBride with Gibbs free energy minimization and a descent Newton-Raphson method, as summarized in Glassman [7]. This method is widely used in commercial tools, such as CHEMKIN®, and free tools such as STANJAN or GasEQ. Without a computer, or a database to support such numerical tools, we still need to estimate flame temperature for educational purposes.

3.2.1 Method of Interpolated Enthalpy

The first hand-calculation method is presented in Kuo [92]. The heat of combustion is distributed in the combustion products. So the products enthalpy at various temperatures are summed to interpolate the flame temperature at which the conservation of enthalpy is met. The implicit assumption is that the flame temperature will be a simple linear function of enthalpy within the targeted temperature range. This process is iterative since the thermochemical data are retrieved more than once. However, the physical meaning of energy conservation is clear in this method. The interpolation process is expressed in Eq. 3.13 and displayed in Fig. 3.2.

$$\frac{T_{AFT} - T_{r1}}{T_{r2} - T_{AFT}} = \frac{\sum H_R - \sum H_p^{T_{r1}}}{\sum H_p^{T_{r2}} - \sum H_R} \quad (3.13)$$

Fig. 3.2 Interpolation of adiabatic flame temperature from known enthalpy points

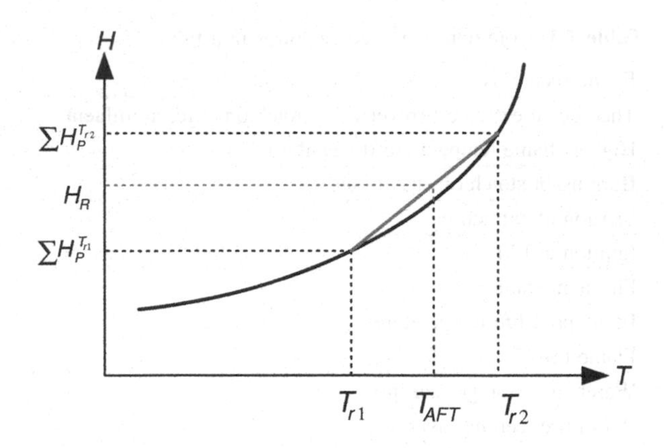

3.2.2 Method of Lump-Sum Specific Heat

Since the above interpolation process is iterative and tedious, a textbook method [91, 93] using lump-sum thermal properties is widely used in both classrooms and fields. The thermal properties at 1000 K are chosen with the implicit assumption that the flame temperature is around 1700 K, so the average temperature is 1000 K between ambient and 1700 K. Thus the non-linearity of thermal properties at high temperature is completely ignored. If the targeted flame temperature is not close to 1700 K, then a try-and-error process or interpolation is needed to converge to the best result.

$$T_{AFT} = T_0 + \frac{\text{production}}{\text{dispersion}} = 298 + \frac{\sum v_i \cdot \Delta H_{C,i}}{\sum v_j \cdot c_{p,j}} \tag{3.14}$$

However, this method is an ideal scheme based on stoichiometric coefficients. If the combustion is incomplete with fuels present in combustion products, it fails to get a meaningful result, since there is nowhere to find the thermal capacity of a fuel.

3.2.3 Method of Accumulative Quenching Potentials

For the adiabatic flame temperature, the energy balance is set up as the following

$$\Delta H_C \cdot x_f = \sum_i Q_{D,i} \cdot \mu_i \cdot E_{air} \tag{3.15}$$

where ΔH_C is the heat of combustion, which can be retrieved from a material table, or estimated from oxygen calorimetry (Eq. 3.12). Q_D is the quenching potential of each species, listed in Table 3.4.

The left-hand side is the heat source, while the right hand side is the summation of energy absorbed by various combustion products, all scaled with the properties

Table 3.4 Thermal properties of common gases

	$\Delta H_p^{1000\,K}$ (kJ/mol)	$\Delta H_p^{2000\,K}$ (kJ/mol)	$\Delta H_p^{3000\,K}$ (kJ/mol)	$C_p^{1000\,K}$ (J/mol K)	Quenching potentials @ 1600 K
Carbon monoxide (CO)	–	–	–	33.2	
Carbon dioxide (CO_2)	33.40	91.44	152.85	54.3	1.75
Water (vapor) (H_2O)	23.12	72.79	126.55	41.2	1.08
Nitrogen (N_2)	21.55	55.98	92.70	32.7	0.99
Oxygen (O_2)	22.71	59.06	98.07	34.9	1.01
Helium (He)/Neo(Ne)/ Argon (Ar)	–	–	–	20.8	0.65

of air (enthalpy of one mole air from ambient to flame temperature). Thus, we can adopt the air correlation in a straightforward way.

$$E_{air} = \frac{\Delta H_C \cdot x_f}{\sum_i Q_{D,i} \cdot \mu_i} = f(T) = 1.4893\mathrm{T}^2 + 29.862\mathrm{T} - 9.381 \tag{3.16}$$

Equation 3.16 can be solved directly, with a positive solution as the adiabatic flame temperature.

$$T_{flame} = \frac{-29.862 + \sqrt{5.9572 \cdot E_{air} + 947.62}}{2.9786} \times 10^3\,\mathrm{K} \tag{3.17}$$

Figure 3.3 shows that this method is closely following the detail methods (GasEQ or Cantera) until close to the stoichiometric state. The deviation of flame temperature near $\phi = 1$ is associated with the reversible reaction near stoichiometry. This method cannot capture the incompleteness of reaction, so it fails at the fuel-rich side

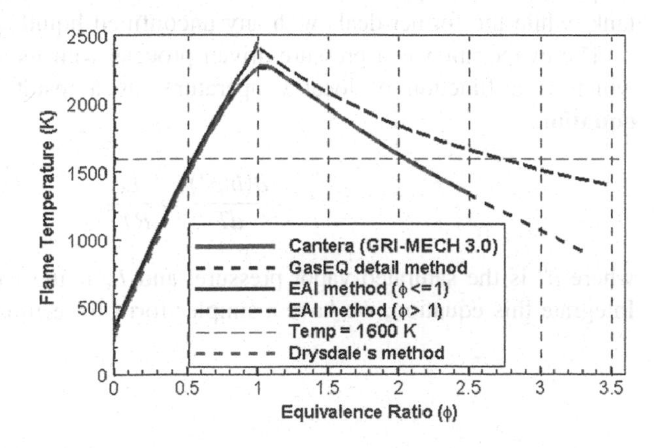

Fig. 3.3 Comparison of estimation result for Propane [16]

of reactions. In fact, as pointed out by Kuo [92], all above estimation methods are based on the assumption that there is no dissociation in combustion products, so we can establish Eqs. 3.13 and 3.14 as representative of the combustion. Lower limits usually occur around 1600 K, within the reliable range of any estimation methods. These methods are valid only when the flame temperature is low. Above this limit, the products will further decompose and dissociate, generating many intermediate products.

3.3 Evaporation Process (Isothermal)

Combustion is essentially a rapid chemical reaction in gas phase. All liquid fuels have to release (vaporize) flammable gases (vapors) before the flaming reaction takes place. This evaporation process can happen with or without fire. When exposed to the ambient atmosphere, any liquid with evaporation will absorb the ambient energy and generate a vapor under a certain vaporization pressure. However, the vapor may not be ignitable upon a pilot flame due to insufficient concentration or insufficient ignition intensity. The ignition of a liquid is a two stage-problem: the vaporization process of liquid fuels and the flammability problem of the liquid vapor. In order to understand the ignition of a liquid fuel, we need to understand the vaporization process first.

3.3.1 Vapor Pressure

Evaporation is a reversible iso-thermal process. When exposed to the open atmosphere, most liquids will evaporate under the normal condition (ambient temperature and pressure), as a result of molecules escaping from the surface to form vapor. If confined, an evaporation equilibrium will be reached where no further net evaporative loss happens at the interface. The latter is a typical case in a gasoline tank, while the former deals with any unconfined liquid spill or liquid pool fires.

The evaporation is a pressure-driven process with its saturated vapor pressure, which is a function of local temperature, as a result of **Clapeyron-Clausius equation**.

$$\frac{d(lnp^0)}{dT} = \frac{L_v}{RT^2} \tag{3.18}$$

where p^0 is the saturated vapor pressure, and L_v is the latent heat of evaporation. Integrate this equation, we have a simpler form for estimating the vapor pressure.

Table 3.5 Vapor pressure constants for some organic compounds [94]

Compound	Formula	E	F	Temperature range (°C)
n-Pentane	n-C_5H_{12}	6595.1	7.4897	−77 to 191
n-Hexane	n-C_6H_{14}	7627.2	7.7171	−54 to 209
Cyciohexane	c-C_6H_{12}	7830.9	7.6621	−45 to 257
n-Oetane	n-C_8H_{18}	9221.0	7.8940	−14 to 281
iso-Octane	C_8H_{18}	8548.0	7.9349	−36 to 99
n-Decane	n-$C_{10}H_{22}$	10912.0	8.2481	17 to 173
n-Dodecane	n-$C_{12}H_{26}$	11857.7	8.1510	48 to 346
Methanol	CH_3OH	8978.8	8.6398	−44 to 224
Ethanol	C_2H_5OH	9673.9	8.8274	−31 to 242
n-Propanol	n-C_3H_7OH	10421.1	8.9373	−15 to 250
Acetone	$(CH_3)_2CO$	7641.5	7.9040	−59 to 214
Methyl ethyl ketone	$CH_3CO \cdot CH_2CH_3$	8149.5	7.9593	−48 to 80
Benzene	C_6H_6	8146.5	7.8337	−37 to 290
Toluene	$C_6H_5CH_3$	8580.5	7.7194	−28 to 31

$$\log_{10} p^0 = -0.2185\frac{E}{T} + F \tag{3.19}$$

where E and F are constants listed in Table 3.5, T is the temperature in Kelvin and p^0 has a unit of mmHg.

3.3.2 *Raoult's Law*

Gases usually appear in the form of a gas mixture, where an "ideal solution" is approximated. For mixtures of ideal gases, **Raoult's law** states that the total pressure is the summation of the component partial pressure of each species. Gas fraction is depending on its partial pressure in the system. Here, the mole (volume) fraction of a gaseous component (i) is defined as

$$x_i = \frac{n_i}{N_i} = \frac{p_i}{P_t} \tag{3.20}$$

Similarly, for the vapors of ideal liquids (satisfying the ideal gas law within a certain temperature range), one can apply Raoult's law in dealing with the partial pressure of vaporization.

$$x_i(liquid) = \frac{n_i}{N_i}(liquid) = \frac{p_i}{P_t}(vapor) \tag{3.21}$$

where p_i is partial pressure of the ith component in solution, P_i is the vapor pressure of the pure component and x_i is its mole fraction in solution.

Raoult's law establishes the relationship of pressure to concentration, so it is important in characterizing the vaporization of a liquid mixture.

3.3.3 The Vaporization of Liquid

From general physics (or thermodynamics), we all know that materials will undergo a phase transition given sufficient energy input. Liquids are typical phase-change materials that are liquids under normal ambient conditions. If the ambient temperature is dropped, the liquid may freeze, like water into ice at 0 °C. For any temperature above the freezing point, the liquid will undergoes the vaporization process, which sets up a concentration profile on top of the liquid surface.

The naturally-happened (constant temperature) vaporization process is demonstrated in the p-v diagrams shown in Fig. 3.4. When the pressure in the liquid phase is dropped, there is a minimum pressure at which the liquid will boil and evaporate. Once all the liquid has turned into the vapor phase, the pressure further drops with vapor expansion. This constant pressure associated with a mixture of both liquid and vapor phase is called ***the saturation vapor pressure***, denoting the state of thermodynamic equilibrium between a liquid and its vapor.

However, most liquid spills on a solid ground will undergo the constant pressure process, as demonstrated in Fig. 3.4b. The super-cooled liquid will absorb the energy from the ambient, raising its surface temperature. When the boiling temperature is reached, the volume begins to expand sharply with surface temperature unchanged. When all the liquid phase is gone, the temperature will rise along with vapor volume expanding. The lower the boiling points, the easier to receive energy from ambient. Therefore, ***the boiling point*** is a critical surface temperature for 100 % vapor at the liquid surface to characterize the ignition potential of a liquid fuel.

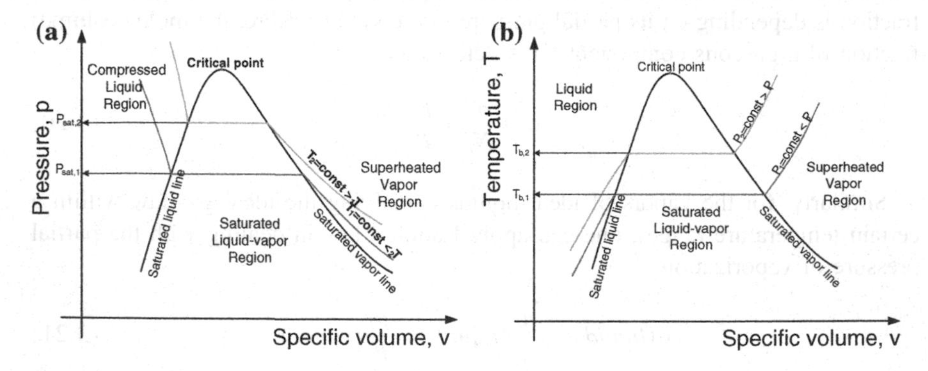

Fig. 3.4 Thermodynamics diagrams for evaporation **a**. p-v liquid-vapor diagram of a pure substance. **b**. T-V diagram of a pure substance undergoing vaporization

The liquid molecules will absorb the energy from ambient, and leave the liquid surface in a gaseous form. The driving force behind this process is the pressure difference between the saturated vapor pressure and ambient pressure. Temperature difference from surface to ambient provides the energy to start the evaporation process. For a pure liquid, the saturated vapor pressure is a function of temperature according to the Clapeyron-Clausius equation.

$$\frac{P_{vs}}{P_0} = \exp\left[\left(\frac{h_{fg}^0}{RT_b}\right)\left(1 - \frac{T_b}{T_s}\right)\right] \quad (3.22)$$

where T_0 and P_0 are the reference temperature and pressure. The subscript s denotes the surface. When T_b is equal to ***the normal boiling point temperature of the liquid***, the reference vapor pressure P_0 will equate to the atmospheric total pressure. h_{fg}^0 is the energy difference from liquid to gas phase, or ***the latent heat of evaporation***. R is a universal gas constant, R = 8.3144 J/mol K. Equation 3.22 determines the vapor pressure at the surface based on the temperature ratio between the ambient temperature and the surface temperature. If the surface temperature T_s approaches T_b, the surface vapor pressure approaches ambient pressure, the fuel concentration at the surface is 100 %, and the liquid begins to boil.

Clapeyron-Clausius equation establishes that the temperature difference will create a pressure difference, which is the driving force behind evaporation or phase transfer. A concentration profile will be established on top of the liquid surface. From Raoult's law, we have the following constitutive relationship as

$$p_F = p_{sat}(T) = x_F p_0 \quad (3.23)$$

Here p_F and p_0 are fuel pressure and ambient pressure respectively. x_F is the volumetric (molar) concentration on the fuel surface. Combined with Clapeyron-Clausius equation (Eq. 3.22), we have the surface concentration established as below.

$$x_F = \frac{p_F}{p_\infty} = \exp\left[\left(-\frac{h_{fg}M_g}{R}\right)\left(\frac{1}{T_s} - \frac{1}{T_b}\right)\right] \quad (3.24)$$

Here $h_{fg} = h_{fg}^0/M_g$ is the mass based latent heat. Since the molecules need energy to leave the liquid surface, latent heat is required to sustain the evaporation, which come from ambient air. Depending on the ambient condition, a freshly spilled liquid pool will have a lower surface temperature during the evaporation process. This temperature drop is sometimes referred as evaporative cooling, as demonstrated in a thermal infared image in Fig. 3.5. A typical application of evaporative cooling is the case when you use saliva on your finger to test the wind direction. If the finger is facing right the incoming wind, the temperature drop is most significant. The temperature sensing cells in your skin will tell you that this is the windward (wind-coming) direction. Table 3.6 lists the thermal properties for common fuels, which are used in estimations between each other.

Fig. 3.5 Due to evaporative cooling, the thermal image of a liquid spilling on parquet floor is darker, showing a lower local temperature [95]

Since evaporative cooling is a diffusion process, we can find the species and temperature profile in Fig. 3.6. At the instant of pouring, the concentration gradient for evaporation is largest, while the temperature difference is smallest. Gradually with evaporative cooling, the surface temperature will be lower than ambient due to this evaporative cooling. A new temperature profile is established when the new thermal equilibrium at the surface is achieved. For estimation purposes, it is commonly assumed that the vapor concentration is linearly distributed, which is implicitly assumed in Sect. 3.4.12.

3.3.4 Flashpoint

Once the concentration profile on top of a liquid surface is established, a natural question is, what is the minimum ambient temperature, which provides barely enough energy for the surface concentration of lower flammability limit, thus a flash of fire is established for short durations? Such a critical temperature is called flashpoint, which has wide applications on the safe handling of liquid fuels. By definition, ***flashpoint*** is the minimum pilot-ignition temperature of a liquid at which sufficient vapor is given off to form an ignitable mixture with the air, near the surface of the liquid or within the vessel used, as determined by the appropriate test procedure and apparatus [96]. Roughly speaking, flashpoint is the minimum ambient temperature that produces the marginally flammable mixture above the liquid surface. Since most fuels are transported and operated under ambient temperatures, the difference between ambient temperature and flashpoint is a good indicator of the energy requirement for initiating a thermal runaway reaction, so representative of the risk of fuel handling. Flash point determinations give rise to qualitative hazard classification systems, the most severe hazard being associated with the liquid with a lowest flash point.

There are several test methods to find the minimum temperature to sustain the ignition. They are shown in Fig. 3.7, which are either close-cup or open-cup devices. For a ***close-cup device***, the fuel is placed in a confined space with sufficient time for the

Table 3.6 Thermal properties of liquid fuels for evaporations [94]

Fuel	Formula	T_L(K)		T_b (K)	T_a (K)	$T_{f,ad}^a$(K)	X_L (%)	h_{fg} (kJ/g)	Δh_c^b (kJ/g)
		Closed	Open						
Methane	CH_4	–	–	111	910	2226	5.3	0.59	50.2
Propane	C_3H_8	–	169	231	723	2334	2.2	0.43	46.4
n-Butane	C_4H_{10}	–	213	273	561	2270	1.9	0.39	45.9
n-Hexane	C_6H_{14}	251	247	342	498	2273	1.2	0.35	45.1
n-Heptane	C_7H_{16}	269	–	371	–	2274	1.2	0.32	44.9
n-Octane	C_8H_{18}	286	–	398	479	2275	0.8	0.30	44.8
n-Decane	$C_{10}H_{22}$	317	–	447	474	2277	0.6	0.28	44.6
Kerosene	$\sim C_{14}H_{30}$	322	–	505	533	–	0.6	0.29	44.0
Benzene	C_6H_6	262	–	353	771	2342	1.2	0.39	40.6
Toluene	C_7H_8	277	280	383	753	2344	1.3	0.36	41.0
Naphthalene	$C_{10}H_8$	352	361	491	799	–	0.9	0.32	40.3
Methanol	CH_3OH	285	289	337	658	–	6.7	1.10	20.8
Ethanol	C_2H_5OH	286	295	351	636	–	3.3	0.84	27.8
n-Butanol	C_4H_9OH	302	316	390	616	–	11.3	0.62	36.1
Formaldehyde	CH_2O	366	–	370	703	–	7.0	0.83	18.7
Acetone	C_3H_6O	255	264	329	738	2121	2.6	0.52	29.1
Gasoline	–	228	–	306	644	–	1.4	0.34	44.1

[a]Based on stoichiometric combustion in air
[b]Based on water and fuel in the gaseous state

equilibrium to be established. If the temperature is low enough, a small flame introduced into the steady-state vapor will not cause ignition, because the surface concentration is smaller than the lower flammable limit. As the fuel temperature is increased, the surface concentration will increase to a certain point that the ignition is established. This minimum temperature to allow the ignition is called **close-cup flashpoint**. Table 3.6 provides a summary of properties of common fuels used for liquid estimations.

In an open-cup configuration, the liquid fuel is placed in an open environment, so there is a concentration gradient near the surface. The concentration of fuel vapors is highest near the surface, and progressively diminishes toward zero away from the surface (as shown in Fig. 3.6). Since the igniter cannot be placed right at the liquid surface, the location of the igniter has a lower vapor concentration and consequently the Open-cup (OC) flashpoint is higher than the closed-cup (CC) value. A correlation is proposed by Factory Mutual [97] for their relationship.

$$T_{FP(OC)} = 1.12 \cdot T_{FP(CC)} + 7.1 \tag{3.25}$$

Theoretically, it is possible to use the LFL value to compute its flashpoint by assuming that the saturation vapor pressure over ambient pressure (surface concentration) is equal to its LFL value (as shown in Fig. 3.8). In reality, the surface heat loss is not negligible and flashpoint determination is device-dependent, so the saturation pressure at LFL is always lower than those expected, which means a

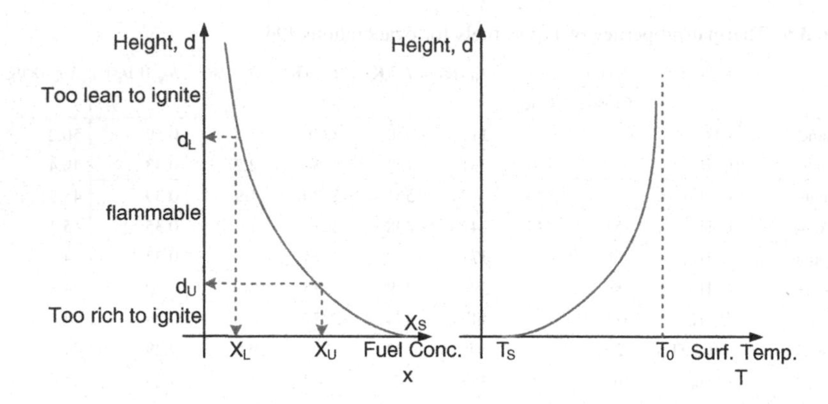

Fig. 3.6 Fuel concentration and temperature profile on top of a flammable fuel surface

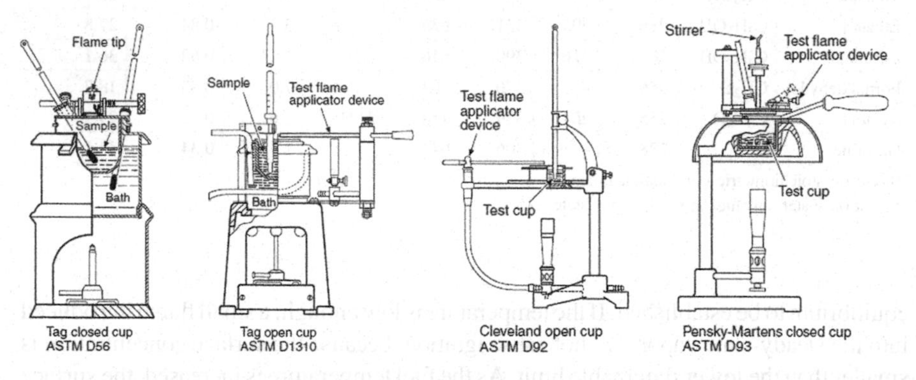

Fig. 3.7 Four common testing devices for flash points [79]

higher fuel concentration than LFL is needed at flashpoint. However, we still assume that the flashpoint is the critical point for the fuel surface to reach lower flammable limit, so we can perform theoretical estimations to understand the evaporation process better.

3.3.5 Estimation of Flashpoints

Since the concentration profile above the liquid surface can be determined theoretically, it is natural to compute flashpoint directly from theory. From the definition, the minimum concentration for flashpoint is the lower flammable limit. However, caution should be made since the heat transfer condition for various liquid fuels are different, so there is some uncertainty involved in using LFL for computing flash point directly. In fact, Kanury [98] found the fuel concentrations at flash point are always above LFL, with some uncertain bias due to heat transfer

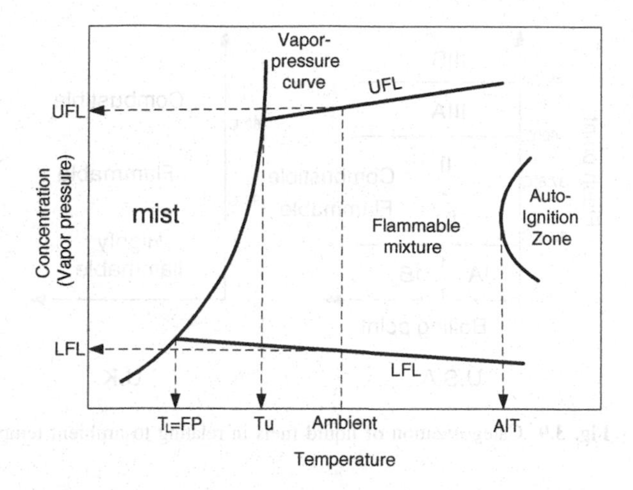

Fig. 3.8 Effect of temperature on liquid vapor pressure

conditions on the fuel surface. Here for educational purposes, we follow Quintiere's lead [91] on calculating the flashpoint theoretically from its lower flammable limit. Using Clapeyron-Clausius equation, the minimum molar concentration at the fuel surface is

$$x_L = e^{-\frac{h_{fg}M_g}{R}\left(\frac{1}{T_S}-\frac{1}{T_b}\right)} \tag{3.26}$$

where x_L is the lower flammable limit of the liquid vapor. Solving this equation, we can find the surface temperature T_S, which is assumed to be at its flashpoint.

3.3.6 Classification of Liquid Fuels

The primary utility of flash points is to classify liquids for safe storage and transportation in various codes and standards. The transportation industry worries about the ambient temperature and its impact on the safety transportation of liquid fuels. NFPA 30 is a primary classification system in US, which defines the following categories based on the flash points relative to the ambient temperature (Table 3.7).

Note such a classification system covers another important parameter: ambient temperature. Here 100 °F (38 °C) is the maximum possible ambient temperature in

Table 3.7 The classification system for liquid fuels in *NFPA30*

Category	NFPA Class	Flash point range
flammable	IA	<73 °F, boiling point <100 °F
	IB	<73 °F, boiling point >100 °F
	IC	>73 °F, <100 °F
combustible	II	>100 °F, <140 °F
	IIIA	>140 °F, <200 °F
	IIIB	>200 °F

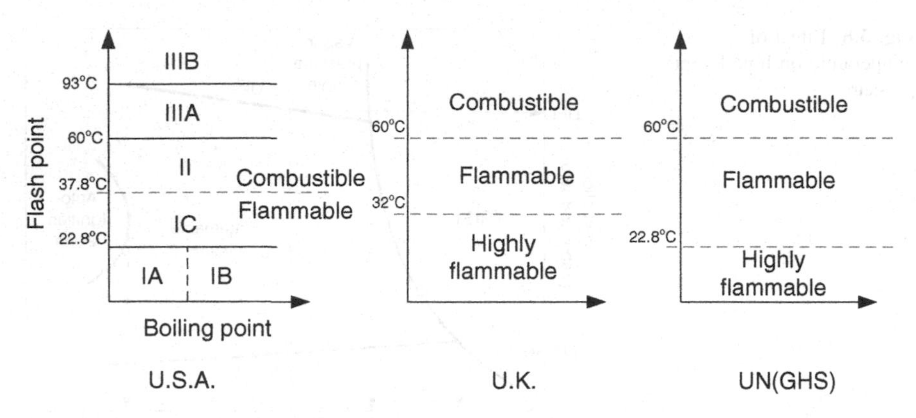

Fig. 3.9 Categorization of liquid fuels in relating to ambient temperature

US. In a northern country, a different and lower temperature system may be adopted for better economy and efficiency in transportation.

Ambient temperature in relative to its flashpoint is a single most important parameter for the fire risk associated with transportation and handling of liquid fuels. A comparison of three classification systems is provided in Fig. 3.9. US system stresses the wide range of ambient temperature, while United Nations (Globally Harmonized System) stresses extreme temperatures. The role of ambient temperature on liquid evaporation and safety is easily recognizable.

3.4 Problems and Solutions

3.4.1 Balance of Stoichiometric Equations

Problem 3.1 Find the stoichiometric coefficients for methane, methanol, ethylene, acetic acid.

Solution:
A spreadsheet with embedded equations is helpful to generate stoichiometric coefficients.

Fuel	Formula	a	b	c	ν_{air}	ν_{CO_2}	ν_{H_2O}	ν_{N_2}
Methane	CH4	1	4	0	2.0	1.0	2.0	7.5
Methanol	CH3OH	1	4	1	1.5	1.0	2.0	5.6
Ethylene	C_2H_4	2	4	0	3.0	2.0	2.0	11.3
Acetic acid	$C_2H_4O_2$	2	4	2	2.0	2.0	2.0	7.5

$$CH_4 + 2(O_2 + 3.76N_2) \rightarrow CO_2 + 2H_2O + 7.52N_2$$
$$CH_4O + 1.5(O_2 + 3.76N_2) \rightarrow CO_2 + 2H_2O + 5.6N_2$$
$$C_2H_4 + 3(O_2 + 3.76N_2) \rightarrow 2CO_2 + 2H_2O + 11.3N_2$$
$$C_2H_4O_2 + 2(O_2 + 3.76N_2) \rightarrow 2CO_2 + 2H_2O + 7.52N_2$$

3.4.2 Non-stoichiometric Equations

Problem 3.2 Balance the reaction for burning propane in air, given the following conditions

a. If the propane molar concentration is 5 %;
b. If the fuel/air ratio is 0.03;

Solution:

a.

$$\phi = \frac{(fuel/air)_{\text{actual}}}{(fuel/air)_{\text{stoich}}} = \frac{(fuel/air)_{\text{actual}}}{r} = \frac{5\,\%/(1-5\,\%)}{1/(5\times 4.76)} = 1.253 > 1$$

$$\mu_{air} = \frac{\nu_{air}}{\phi} = \frac{1-x_f}{4.76\cdot x_f} = \frac{1-0.05}{4.76\times 0.05} = 3.99$$

$$\mu_{CO_2} = (1-\mu_f)\cdot a = (1-0.202)\times 3 = 2.394$$

$$\mu_{H_2O} = (1-\mu_f)\cdot\frac{b}{2} = (1-0.202)\times\frac{8}{2} = 3.192$$

$$\mu_N = 3.76\cdot\mu_{air} + (1-\mu_f)\cdot\frac{d}{2} = 15.0$$

$$\mu_O = 0$$

$$\mu_f = 1 - \frac{\mu_{air}}{\nu_{air}} = 1 - \frac{3.99}{5} = 0.202$$

Therefore, the reaction equation for 5 % fuel in mixture is balanced as:

$$C_3H_8 + 3.99(O_2 + 3.773N_2) \rightarrow 2.394CO_2 + 3.192H_2O + 0.202C_3H_8 + 15.0N_2$$

b.

$$\phi = \frac{(fuel/air)_{\text{actual}}}{r} = \frac{0.03}{1/(5\times 4.76)} = 0.714 < 1$$

$$\mu_{air} = \frac{\nu_{air}}{\phi} = \frac{1-x_f}{4.76\cdot x_f} = \frac{5}{0.714} = 7$$

$$\mu_{CO_2} = a = 3$$

$$\mu_{H_2O} = \frac{b}{2} = 4$$

$$\mu_N = 3.76\cdot\mu_{air} + \frac{d}{2} = 26.32$$

$$\mu_O = \mu_{air} - a - \frac{b}{2} = 7 - 3 - 2 = 2$$

$$\mu_f = 0$$

Therefore, the reaction for fuel/air ratio of 0.03 is

$$C_3H_8 + 7(O_2 + 3.76N_2) \rightarrow 3CO_2 + 4H_2O + 2O_2 + 26.32N_2$$

3.4.3 Heat of Formation

Problem 3.3 Compute the heat of formation of Toluene, if its heat of combustion is 3733.11 kJ/mol.

Solution:
First, we need to find the stoichiometric reaction for Heptane.

$$C_7H_8 + \nu_{air}(O_2 + 3.76N_2) \rightarrow \nu_{CO_2}CO_2 + \nu_{H_2O}H_2O + \nu_{N_2}N_2$$

Applying Eq. 3.2 for stoichiometric coefficients, we have $\nu_{air} = 9$, $\nu_{CO_2} = 7$, $\nu_{H_2O} = 4$, and $\nu_{N_2} = 33.96$. Thus, we have the balanced equation as

$$C_7H_8 + 9(O_2 + 3.76N_2) \rightarrow 7CO_2 + 4H_2O + 33.96N_2$$

Second, we need to find the heat of formation for all reactants and products. From Table 3.1, we can find the heat of formation for n-Heptane, Oxygen, Nitrogen, CO_2, H_2O are −187.8, 0, 0, −393.51, −241.83 (vapor), respectively.

Then,

$$\begin{aligned}\Delta H_C &= \sum \nu_i \cdot \Delta H^0_{f,i}\Big|_{\text{reactants}} - \sum \nu_j \cdot \Delta H^0_{f,j}\Big|_{\text{products}}\\ &= 1 \times \Delta H_f + 9 \times (0 + 3.76 \times 0)\\ &\quad - [7 \times (-110.5) + 4 \times (-241.8) + 33.96 \times 0]\\ &= 3733.11\ \text{kJ/mol}\end{aligned}$$

So $\Delta H_f = 11.22$ kJ/mol

Problem 3.4 Determine by calculation the enthalpy of formation in kJ/mol of Methane CH_4 given its heat of combustion is 50.0 kJ/g @ 25 °C, and the heats of formation of carbon dioxide and water vapor are respectively: −393.5 and −241.8 kJ/mol.

Solution:

$$
\begin{aligned}
&CH_4 + 2O_2 \rightarrow CO_2 + 2H_2O \\
&H_1 = \Delta h_{f,CH_4} \\
&H_2 = (-393.5) + (-241.8) \times 2 = -877.1 \text{ kJ/mol} \\
&\Delta h_C(CH_4) = 50 \text{ kJ/g} \times 16 \text{ g/mol} = 800 \text{ kJ/mol} = -877.1 - \Delta h_{f,CH_4} \\
&\Delta h_{f,CH_4} = -77.1 \text{ kJ/mol}
\end{aligned}
$$

3.4.4 Heat of Combustion

Problem 3.5 Compute and estimate the heat of combustion of n-Heptane.

Solution:
First, we need to find the stoichiometric reaction for Heptane.

$$C_7H_{16} + 11(O_2 + 3.76N_2) \rightarrow 7CO_2 + 8H_2O + 41.36\,N_2$$

Second, we need to find the heat of formation for all reactants and products. From Table 3.2, we can find the heat of formation for n-Heptane, Oxygen, Nitrogen, CO_2, H_2O are −187.8, 0, 0, −393.51, −241.83 (liquid), respectively.

Then,

$$
\begin{aligned}
\Delta H_C &= \sum v_i \cdot \Delta H^0_{f,i}\Big|_{\text{reactants}} - \sum v_j \cdot \Delta H^0_{f,j}\Big|_{\text{products}} \\
&= 1 \times (-224.4) + 9 \times (0 + 3.76 \times 0) - [7 \times (-393.51) + 8 \times (-241.8) + 33.96 \times 0] \\
&= 4464.8 \text{ kJ/mol}
\end{aligned}
$$

$$\text{Or } \Delta H_C = \frac{4464.8 \text{ kJ/mol}}{100 \text{ g/mol}} = 44.6 \text{ kJ/g}$$

By oxygen calorimetry, the heat of combustion for this reaction is estimated as

$$\Delta H_C = 0.414 \times 11 = 4554 \text{ kJ/mol}$$

Or by oxygen mass consumption: $\Delta H_C = 13.1 \times 11 \times 32 = 4611.2$ kJ/mol

Problem 3.6 Find the heat of combustion for the following reactions

a. $CH_4 + 2O_2 \rightarrow CO_2 + 2H_2O(l)$
b. $CH_4 + 2O_2 \rightarrow CO_2 + 2H_2O(g)$

Solution:

$$CH_4{+}2O_2 \rightarrow CO_2 + 2H_2O(l)$$
$$-74.870{-}393.51(-285.83) \times 2$$
$$\Delta H_C = -74.87 - (-393.51) - (-285.83) \times 2 = 890.3 \text{ kJ/mol}$$
$$CH_4 + 2O_2 \rightarrow CO_2 + 2H_2O(g)$$
$$CH_4{+}2O_2 \rightarrow CO_2 + 2H_2O(g)$$
$$-74.870{-}393.51(-241.83) \times 2$$
$$\Delta H_C = -74.87 - (-393.51) - (-241.83) \times 2 = 802.3 \text{ kJ/mol}$$

Problem 3.7 Compute the heat of combustion per gmole of acetonitrile at 25 °C. Acetonitrile (C_2H_3N) burns to form hydrogen cyanide (HCN), carbon dioxide and water vapor with thermal properties supplied below [91].

Species	Heat of formation (kcal/gmol)
Hydrogen cyanide	32.3
Acetonitrile	21.0
Water vapor	−57.8
Carbon dioxide	−94.1
Oxygen	0

Assume constant and equal specific heats of constant pressure and constant volume of 1.2 and 1.0 kJ/kg K.

Solution:

$$C_2H_3N + 1.5O_2 \rightarrow CO_2 + H_2O + HCN$$
$$H_1 = 21.0 \text{ kcal/mol}$$
$$H_2 = -94.1 + (-57.8) + 32.3 = -119.6 \text{ kcal/mol}$$
$$\Delta h_C(C_2H_3N) = H_2 - H_1 = -119.6 - 21 = -140.6 \text{ kcal/mol}$$

Problem 3.7 Answer the following questions.

(a) Calculate the enthalpy change (20 °C) in the oxidation of **n-Pentane** to carbon monoxide and water $C_5H_{12} + 5.5O_2 \rightarrow 5CO + 6H_2O$, Express the result in terms of energy released per gram of n-Pentane burnt; express the result in terms of energy released per gram of air consumed.

(b) the products of the partial combustion of **n-Pentane** (C_5H_{12}) were found to contain CO_2 and CO in the ratio of 4:1. What is the actual heat released per mole of n-pentane burnt if the only other product is H_2O?

Solution:

(a) $\Delta H = H_2 - H_1$

$$H_2 = \left(\Delta H_{f,CO_2} + c_{p,CO_2}\Delta T\right) \times 5 + \left(\Delta H_{f,H_2O} + c_{p,H_2O}\Delta T\right) \times 6$$
$$= [(-110.5) \times 5 + 0.0332 \times (-5)] \times 5 + [(-241.8) \times 1 + 0.0412 \times (-5)] \times 6$$
$$= -2005.04 \text{ kJ/mol}$$

$$H_2 = \left(\Delta H_{f,C_5H_{12}} + c_{p,C_5H_{12}}\Delta T\right) \times 1 + \left(\Delta H_{f,O_2} + c_{p,O_2}\Delta T\right) \times 5.5$$
$$= [(-159.3) \times 1 + 0.06 \times (-5)] \times 1 + [0 \times 1 + 0.0349 \times (-5)] \times 5.5$$
$$= -160.60 \text{ kJ/mol}$$

$$\Delta H_C = H_2 - H_1 = -2005.04 + 160.6 = 1844.44 \text{ kJ/mol}$$

$$\text{Or } \Delta H_C = \frac{1844.44\,\text{kJ/mol}}{72\,\text{g/mol}} = 25.62\,\text{kJ/g}$$

(b) $C_5H_{12} + aO_2 \rightarrow bH_2O + c(CO + 4CO_2)$

For the conservation of H atoms $2b = 12 \rightarrow b = 6$

For the conservation of C atoms $5 = c(1 + 4) \rightarrow c = 1$

For the conservation of O atoms $2a = 6 + 1 + 8 = 15 \rightarrow a = 7.5$

$$C_5H_{12} + 7.5O_2 \rightarrow 6H_2O + 1(CO + 4CO_2)$$
$$H_2 = (-241.8) \times 6 + [(-110.5) + (-393.5) \times 4] \times 1 = -3135.3 \text{ kJ/mol}$$
$$H_1 = (-159.3) \times 1 = -159.3 \text{ kJ/mol}$$
$$Q = H_2 - H_1 = -3135.3 - (159.3) = -2976 \text{ kJ/mol}$$

3.4.5 Flame Temperature by Interpolation of Enthalpy

Problem 3.8 Methane (CH_4) is burned in air at a constant pressure with 298 K and 1 atm. Determine the adiabatic flame temperature for this condition assuming complete combustion.

Solution:

(1) Balance Chemical Equation

$$CH_4 + 2(O_2 + 3.76\,N_2) \rightarrow CO_2 + 2H_2O + 7.52\,N_2$$

(2) Applying energy balance and the adiabatic assumption, we have

$$\sum H_R = \sum H_P$$

(3) Now determine the enthalpy of reactants

$$\sum H_R = \sum \mu_i \cdot \Delta H_C = 802.3 \text{ kJ/mol}$$

(4) Assume the flame happens at 1000 K, now determine the enthalpies of products

$$\sum H_P = 33.40 + 2 \times 23.12 + 7.52 \times 21.40 = 240.57 \text{ kJ/mol}$$

This enthalpy sum is far less than the enthalpy of reactants.
Now assume the flame happens around 2000 K,

$$\sum H_P = 91.44 + 2 \times 72.79 + 7.52 \times 55.98 = 657.97 \text{ kJ/mol}$$

It is still not enough to match the enthalpy of reactants.
Now assume the flame happens around 3000 K,

$$\sum H_P = 152.85 + 2 \times 126.55 + 7.52 \times 92.98 = 1103.02 \text{ kJ/mol}$$

(5) This time the products have more energy than reactants, so the flame temperature should be somewhere between 2000 and 3000 K, which fine-tuned through interpolation.

$$\frac{T_{AFT} - 2000}{3000 - T_{AFT}} = \frac{802.3 - 657.97}{1103.02 - 803.3}$$

Solving this interpolation, we have $T_{AFT} = 2324.3$ K

3.4.6 Flame Temperature by Lump-Sum Specific Heat

Problem 3.9 Calculate the adiabatic flame temperature (at constant pressure) for ethane C_2H_6 in air.

a. at the stoichiometric mixture condition.
b. at the lower flammable limit (5.3 %)

Solution:
First, we need to establish the stoichiometric reaction as below.

$$C_2H_6 + 3.5(O_2 + 3.76N_2) --> 2CO_2 + 3H_2O + 13.16N_2$$

Second, we will have the lumped energy and specific terms.

$$\Delta H_c = 1423\,\text{kJ/mol}$$

$$\sum n_i c_{p,i} = 2 \times 54.3 + 3 \times 41.2 + 13.16 \times 32.7 = 662.5\,\text{J/mol K}$$

Finally, the flame temperature is estimated using Eq. (3.10).

$$T_{AFT} = 298 + \frac{1{,}423{,}000}{662.5} = 2446\,\text{K}$$

For the second part of the problem, we need to establish the fuel-lean reaction using the lower flammability limits (3.0 % for ethane).

First, find the air/fuel ration from the lower flammability limit.

$$(1 - 3.0\,\%)/3.0\,\% = 32.33$$

Next find the oxygen coefficient (or oxygen/fuel molecular ratio) in the reaction.

$$32.33/4.76 = 6.79$$

Third, a reaction is established as below.

$$C_2H_6 + 6.79O_2 + 25.54N_2 --> 2CO_2 + 3H_2O + 3.29O_2 + 25.54N_2$$

Then, the lumped specific heat is below.

$$\sum n_i c_{p,i} = 2 \times 54.3 + 3 \times 41.2 + 25.54 \times 32.7 + 3.29 \times 34.9 = 1182.2\,\text{J/mol K}$$

Finally, the adiabatic flame temperature is found using Eq. (3.10).

$$T_{AFT} = 298 + \frac{1{,}423{,}000}{1182.2} = 1502\,\text{K}$$

Problem 3.10 Calculate the adiabatic flame temperature (at constant pressure) for acetylene burning in air.

(a) At the stoichiometric mixture condition.
(b) At the lower flammable limit.

Solution:

(a) $C_2H_2 + 2.5O_2 + 9.4N_2 \rightarrow 2CO_2 + H_2O + 9.4N_2$

$$T_{AFT} = T_0 + \frac{\sum v_i \cdot \Delta H_c}{\sum n_i, c_{p,i}} = 298 + \frac{1255500}{2(54.3) + 1(41.2) + 9.4(32.7)} = 3044.18\,\text{K}$$

(b) At lower flammable limit of 2.5 %

Air/fuel ratio is $\frac{1-0.025}{0.025} = 39$
Oxygen/fuel molar ratio is $\frac{39}{4.76} = 8.193$

$C_2H_2 + 8.193O_2 + 30.806N_2 \rightarrow 2CO_2 + H_2O + 5.693O_2 + 30.806N_2$

$$T_{AFT} = T_0 + \frac{\sum v_i \cdot \Delta H_c}{\sum n_i, c_{p,i}} = 298 + \frac{1255500}{2(54.3) + 1(41.2) + 5.693(34.9) + 30.806(32.7)} = 1224\,\text{K}$$

Problem 3.11 Calculate the adiabatic flame temperature (at constant pressure) for propane in air.

a. At the stoichiometric mixture condition.
b. At the lower flammable limit

Solution:

$$C_3H_8 + v_{air}(O_2 + 3.76N_2) \rightarrow v_{CO_2}CO_2 + v_{H_2O}H_2O + v_{N_2}N_2$$

$$v_{air} = 3 + \frac{8}{4} = 5;\; v_{CO_2} = 3;\quad v_{H_2O} = \frac{8}{2} = 4$$

Balanced Chemical Equation:

$$C_3H_8 + 5O_2 + 18.8N_2 \rightarrow 3CO_2 + 4H_2O + 18.8N_2$$

(a) $$T_{AFT} = T_0 + \frac{\sum v_i \cdot \Delta H_c}{\sum n_i, c_{p,i}} = 298 + \frac{2043100}{3(54.3) + 4(41.2) + 18.8(32.7)} = 2465.8\,\text{K}$$

(b) LFL = 2.1 %

Air/Fuel Ratio = $\frac{1-0.021}{0.021} = 46.62$
Oxygen Coefficient = $\frac{46.62}{4.76} = 9.794$

$$C_3H_8 + 9.794O_2 + 36.825N_2 \rightarrow 3CO_2 + 4H_2O + 4.794O_2 + 36.825N_2$$

$$T_{AFT} = T_0 + \frac{\sum v_i \cdot \Delta H_c}{\sum n_i, c_{p,i}} = 298 + \frac{2043100}{3(54.3) + 4(41.2) + 4.794(34.9) + 36.825(32.7)} = 1500.4\,\text{K}$$

3.4.7 Flame Temperature by Quenching Potentials

Problem 3.12 Methane (CH_4) is burned in air at a constant pressure with 298 K and 1 atm. Determine the adiabatic flame temperature for this condition assuming complete combustion.

Solution:

(1) Balance Chemical Equation

$$CH_4 + 2(O_2 + 3.76\,N_2) \rightarrow CO_2 + 2H_2O + 7.52\,N_2$$

(2) Applying energy balance and the adiabatic assumption, we have

$$E_{air} = \frac{\Delta H_C \cdot \mu_f}{\sum_i EAI_i \cdot \mu_i} = \frac{802.3 \times 1}{1.63 \times 1 + 2 \times 1.08 + 7.52 \times 0.99} = 71.41\ \text{kJ/mol}$$

01. Now solve Eq. (3.13) directly, we have

$$T_{AFT} = \frac{-29.862 + \sqrt{5.9572 \cdot 71.41 + 947.62}}{2.9786} \times 10^3 = 2414.7\,\text{K}$$

Problem 3.13 Find the adiabatic flame temperature for *Pentane* at LFL (C_5H_{12}, LFL = 0.014).

Solution:

a. Stoichiometric Chemical Equation

$$C_5H_{12} + 8(O_2 + 3.76\,N_2) \rightarrow 5\ CO_2 + 6\ H_2O + 30.08\,N_2$$

b. Non-stoichiometric reaction @ LFL ($x_L = 0.014$)

$$\phi = \frac{(fuel/air)_{\text{actual}}}{(fuel/air)_{\text{stoich}}} = \frac{1.4\,\%/(1 - 1.4\,\%)}{1/(8 \times 4.76)} = 0.54 < 1$$

$$\mu_{air} = \frac{v_{air}}{\phi} = \frac{8}{\phi} = 14.8$$

$$\mu_{CO_2} = a = 5$$

$$\mu_{H_2O} = \frac{b}{2} = 6$$

$$\mu_N = 3.76 \cdot \mu_{air} + \frac{d}{2} = 55.6$$

$$\mu_O = \mu_{air} - a - \frac{b}{4} + \frac{c}{2} = 14.8 - 5 - \frac{12}{4} + 0 = 6.8$$

$$\mu_f = 0$$

c. So the reaction equation for 3 % fuel in mixture is balanced as:

$$C_5H_{12} + 14.8\,(O_2 + 3.76N_2) \rightarrow 5CO_2 + 6H_2\,O + 6.8O_2 + 55.6\,N_2$$

d. Since $\Delta H_C = 3245.31$ kJ/mol, applying energy balance and the adiabatic assumption, we have

$$E_{air} = \frac{\Delta H_C \cdot \mu_f}{\sum_i EAI_i \cdot \mu_i} = \frac{3245.31 \times 1}{1.63 \times 5 + 6 \times 1.08 + 6.8 \times 1.01 + 55.6 \times 0.99}$$
$$= 42.4 \text{ kJ/mol}$$

e. Now solve the energy equation directly, we have

$$T_{AFT} = \frac{-29.862 + \sqrt{5.9572 \times 42.4 + 947.62}}{2.9786} \times 10^3 = 1605.4\text{K}$$

f. Or using lump-sum specific heat

$$T_{AFT} = 298 + \frac{\Delta H_C \cdot \mu_f}{\sum_i C_{p,i} \cdot \mu_i}$$
$$= 298 + \frac{3245310}{54.3 \times 5 + 6 \times 41.2 + 6.8 \times 34.9 + 55.6 \times 32.7} = 1558.7\text{K}$$

3.4.8 Suppression Modeling by Flame Temperature

Problem 3.15 Calculate the lower flammability limit of propane in a mixture of (a) 21 % oxygen + 79 % Nitrogen, air; (b) 21 % oxygen + 79 % helium, and (c) 21 % oxygen + 79 % carbon dioxide, assuming a limiting adiabatic flame temperature of 1600 K. (Initial temperature 20 °C) [94].

Solution:

(a) Stoichiometric Chemical Equation $C_3H_8 + 5(O_2 + 3.76\,N_2) \rightarrow 3\,CO_2 + 4\,H_2O + 18.8\,N_2$

(b) Determine the coefficients

$$\mu_{air} = \frac{v_{air}}{\phi} = \frac{5}{\phi} = \frac{(1 - x_L)}{4.76 \cdot x_L}$$

$$\mu_{CO_2} = a = 3$$

$$\mu_{H_2O} = \frac{b}{2} = 4$$

$$\mu_N = 3.76 \cdot \mu_{air} + \frac{d}{2} = 0.79 \cdot (1 - x_L)/x_L$$

$$\mu_O = \mu_{air} - a\frac{b}{4} + \frac{c}{2} = \frac{(1 - x_L)}{4.76 \cdot x_L} - 5$$

$$\mu_f = 0$$

(c) Using lump-sum specific heat, for nitrogen

$$T_{AFT} = 298 + \frac{\Delta H_C \cdot \mu_f}{\sum_i C_{p,i} \cdot \mu_i} = 298 + \frac{2043100}{54.3 \times 3 + 4 \times 41.2 + \left(0.21\frac{1-x_L}{x_L} - 5\right) \times 34.9 + \frac{0.79(1-x_L)}{x_L} \times 32.7} = 1600\,\text{K}$$

$x_L = 0.023$

(d) Using lump-sum specific heat, for helium

$$T_{AFT} = 298 + \frac{\Delta H_C \cdot \mu_f}{\sum_i C_{p,i} \cdot \mu_i} = 298 + \frac{2043100}{54.3 \times 3 + 4 \times 41.2 + \left(0.21\frac{1-x_L}{x_L} - 5\right) \times 34.9 + \frac{0.79(1-x_L)}{x_L} \times 20.8} = 1600\,\text{K}$$

$x_L = 0.017$

(e) Using lump-sum specific heat, for carbon dioxide.

$$T_{AFT} = 298 + \frac{\Delta H_C \cdot \mu_f}{\sum_i C_{p,i} \cdot \mu_i} = 298 + \frac{2043100}{54.3 \times 3 + 4 \times 41.2 + \left(0.21\frac{1-x_L}{x_L} - 5\right) \times 34.9 + \frac{0.79(1-x_L)}{x_L} \times 54.3} = 1600\,\text{K}$$

$x_L = 0.034$

3.4.9 Flashpoint by Integrated Clasius-Claperon Equation

Problem 3.15 If a n-Hexane liquid is spilled at STP(20 °C (293.15 K, 68 °F) and 1 atm). Can you ignite this vapor at the surface?

Solution:
The LFL and UFL of n-Hexane are 1.2 and 7.4 % respectively. This translate to a vapor pressure of $1.2\,\% \times 760 = 9.12$ mmHg and $7.4\,\% \times 760 = 56.24$ mmHg. The calculation shows the vapor pressure is 107 mmHg, well above the upper limit. So the fuel concentration at the surface is too rich to ignite. However, because of the diffusion process in evaporation, there is a flammable zone above the surface, so you can still ignite the liquid, if the flammable zone is ignited.

$$\log P = -0.2185\frac{7627.2}{293.15} + 7.7171 = 2.032$$
$$p^0 = 10^{2.032} = 107.7 \text{ mmHg}$$

Problem 3.16 Calculate the temperature at which the vapor pressure of n-Decane corresponds to the lower flammability limit for n-Decane vapor. Assume that this vapor pressure is 0.75 % by volume and that the atmospheric pressure is 760 mmHg.

Solution:
The vapor pressure corresponding to the lower flammability limit of n-hexane is 0.75 % of 760 mmHg, or 5.7 mmHg. Then $\log_{10} p = \log_{10}(5.7) = 0.756$

Thus

$$0.756 = -0.2185\frac{10912}{T} + 8.24809$$
$$4.836T = 1666.543$$
$$T = 344.8\,\text{K} = 71.6\,°\text{C}$$

The closed cup flashpoint of n-Decane is given as 317 K, or 44 °C.

Problem 3.17 Calculate the temperature at which the vapor pressure of n-Decane corresponds to a stoichiometric vapor-air mixture.

$$C_{10}H_{22} + 14.5(O_2 + 3.76N_2) \rightarrow \text{stoichiometric}$$

Solution:

$$x_{st} = \frac{1}{1 + 14.5 \times 4.76} = 0.014185$$

$$M = 142 \text{ g/mol}, \; h_{fg} = 280\,\text{J/g}, \; T_b = 447\,\text{K}$$

$$x_{st} = e^{-\frac{h_{fg}M}{R}\left(\frac{1}{T} - \frac{1}{T_b}\right)}$$

$$\ln(0.014185) = -\frac{280\,\text{J/g} \times 142\,\text{g/mol}}{8.3144}\left(\frac{1}{T} - \frac{1}{447}\right)$$

$$T = 319.8\,\text{K}$$

Problem 3.18 Estimate the minimum piloted ignition temperature for **methanol** in air, which is at 25 °C and 1 atm.

Solution:

$$x_L = 0.067, \; \text{T}_\text{b} = 337\,\text{K}, \; \text{h}_\text{fg} = 1100\,\text{J/g}, \; \text{M} = 32/\text{mol}$$

$$0.067 = \text{e}^{-\frac{32 \times 1100}{8.315}\left(\frac{1}{T} - \frac{1}{337}\right)} \rightarrow T = 277\,\text{K} = 4\,°C$$

As a comparison, the published flash point open cup value is 285 K, while the close-cup value is 289 K.

Problem 3.19 Calculate the temperature at which the vapor pressure of ***acetone*** corresponds to 0.036 atm.

Solution:

$$T_b = 56\,°\text{C}; \; h_{fg} = 540 \text{ J/g}; \; MW = 58 \text{ g/mol}$$

$$\ln(0.036) = -\frac{520 \text{ J/g} \times 58 \text{ g/mol}}{8.3144}\left(\frac{1}{T} - \frac{1}{56 + 273.15}\right) \rightarrow T = -20.3°\text{C}$$

As a point of reference, the published close-cup flashpoint for acetone is −18 °C.

Problem 3.20 Calculate the temperature at which the vapor pressure of ***Heptane*** (C_7H_{16}) is barely flammable (LFL) at the surface. Note $h_{fg} = 320$ J/g.

Solution:

$$T_b = 98°C = 371\,\mathrm{K};\; MW = 100\,\mathrm{g/mol}$$

$$x_L = e^{-\frac{h_{fg}M}{R}\left(\frac{1}{T_L}-\frac{1}{T_b}\right)}$$

$$\ln(0.012) = -\frac{320\,\mathrm{J/g} \times 100\,\mathrm{g/mol}}{8.3144}\left(\frac{1}{T_L}-\frac{1}{371}\right) \rightarrow T_L = 260\,\mathrm{K} = -13\,°\mathrm{C}$$

The listed FP value is −4 °C for Heptane.

3.4.10 Vapor Pressure of a Mixture

Problem 3.21 If a n-Hexane liquid is spilled at STP (20 °C (293.15 K, 68 °F) and 1 atm). Find the vapor pressure of n-Hexane for this spill.

Solution:

$$\log_{10} p^0 = -0.2185\frac{E}{T} + F = -0.2185 \times \frac{7627.2}{293.15} + 7.7171 = 2.03$$
$$p^0 = 107\,\mathrm{mmHg}$$

Problem 3.22 Calculate the vapor pressures of *n*-hexane and *n*-Decane above a mixture at 25 °C containing 2 % *n*-C_6H_{14} + 98 % *n*-$C_{10}H_{22}$, by volume. Assume that the densities of pure *n*-hexane and *n*-Decane are 660 and 730 kg/m^3, respectively and that the liquids behave ideally.

Solution:
Note the mixture is presented as volume ratio, which needs to be converted to molar concentration for apply Raoult's rule.

a. Find mole fraction of each component

$$n_{C_6H_{14}} = \frac{0.02 \times 660/0.086}{0.02 \times 660/0.086 + 0.98 \times 730/0.142} = 0.02957$$

$$n_{C_{10}H_{22}} = \frac{0.98 \times 730/0.142}{0.02 \times 660/0.086 + 0.99 \times 730/0.142} = 0.9704$$

b. Find the saturated vapor pressure of the fuels

$$\text{For } C_6H_{14},\ \log_{10} p^o = -0.2185 \frac{7627.2}{273+25} + 7.7171 = 2.125 \rightarrow p^o = 133.3 \text{ mmHg}$$

$$\text{For } C_{10}H_{22},\ \log_{10} p^o = -0.2185 \frac{10912}{273+25} + 8.2481 = 0.247 \rightarrow p^o = 1.767 \text{ mmHg}$$

c. Applying Raoult's rule

$$\text{For hexane, } p = n_{C_6H_{14}} \cdot p^o = 0.02957 \times 133.3 = 3.94 \text{ mmHg}$$

$$\text{For n-Decane, } p = n_{C_{106}H_{22}} \cdot p^o = 0.9704 \times 1.767 = 1.715 \text{ mmHg}$$

3.4.11 Flashpoint of a Binary Mixture

Problem 3.23 Determine by calculation whether **n-Decane** containing 1 % n-pentane (by volume) would be classified as a Class 1C or a Class II flammable liquid according to the NFPA Standard (This is equivalent to posing the question, is the flashpoint above or below 37.8 °C?)

Solution:

a. Collect Initial Inputs:

	Pentane (C_5H_{12})	n-Decane ($C_{10}H_{22}$)
Density (kg/m^3)	626	730
Volume fraction	0.01	0.99
lower flammable limits	0.014	0.0075
Molecular weight (g/mol)	72	142

b. Find mole fraction of each component

$$x_{C_5H_{12}} = \frac{0.01 \times 626/0.072}{0.01 \times 626/0.072 + 0.99 \times 730/0.142} = 0.0168$$

$$n_{C_{10}H_{22}} = \frac{0.99 \times 730/0.142}{0.01 \times 626/0.072 + 0.99 \times 730/0.142} = 0.9832$$

c. Find saturation pressure of each component

$$\log p_{C_5H_{12}} = -0.2185 \times \frac{6595.1}{273.2 + 37.8} + 7.4897 = 2.856 \rightarrow p_{C_{11}H_{24}} = 718 \text{ mmHg}$$

$$\log p_{C_{10}H_{22}} = -0.2185 \times \frac{10912.0}{273.2 + 37.8} + 8.2481 = 0.581 \rightarrow p_{C_{10}H_{22}} = 3.816 \text{ mmHg}$$

d. Find the vapor partial pressure in the mixture

$$p_{C_5H_{12}} = p^0_{C_5H_{12}} \cdot n_{C_5H_{12}} = 0.0168 \times 718 = 12.06\,\text{mmHg}$$

$$p_{C_{10}H_{22}} = p^0_{C_{10}H_{22}} \cdot n_{C_{10}H_{22}} = 0.9832 \times 3.816 = 3.7519\,\text{mmHg}$$

e. Apply Le Chatelier's rule

$$\frac{12.06/760}{0.014} + \frac{3.7519/760}{0.0075} = 1.79 > 1$$

So the lower limit has been exceeded. Thus the liquid mixture has a flashpoint below 37.8 °C, or the mixture is definitely not class II liquid (see Fig. 3.9).

Problem 3.24 N-Dodecane has a closed cup flashpoint of 74 °C. What percentage by volume of n-hexane would be sufficient to give a mixture with a flashpoint of 32 °C ?

Solution:

$$\log p_{C_{12}H_{26}} = -0.2185 \times \frac{11857.7}{305} + 8.1510 = -0.3438 \rightarrow p_{C_{12}H_{26}} = 0.4531\ \text{mmHg}$$

$$\log p_{C_6H_{14}} = -0.2185 \times \frac{7627.2}{305} + 7.7171 = 2.253 \rightarrow p_{C_6H_{14}} = 178.9\ \text{mmHg}$$

$$\frac{100 \times 178.9n/760}{0.95} + \frac{100 \times 0.4531(1-n)/760}{0.6} = 1$$

$$19.619n + 0.099 - 0.099n = 1$$

$$19.52n = 0.9006$$

$$n = 0.0461$$

$$n_{C_6H_{14}} = \frac{m \times 654.8/0.086}{n \times 654.8/0.086 + (1-m) \times 749/0.170} = 0.0461$$

Solving this equation, the volume fraction of hexane is m = 0.0271 or 2.7 %.

3.4.12 Evaporation Profiles on Liquid Surface

Problem 3.25 N-Octane spills on a hot pavement during a summer day. The pavement is 40 °C and heats the octane to this temperature.

Data: ambient condition: 33 °C, 760 mmHg pressure
Density of air = 1.2 kg/m^3
Specific heat of air = 1.04 J/g K
Boiling point = 125.6 °C

Heat of vaporization = 0.305 kJ/g
Specific heat (gas) = 1.67 J/g.K
Density (liquid) = 705 kg/m3
Lower flammability limit in air = 0.95 %
Upper flammability limit in air = 3.20 %

a. Determine the molar concentration of the *octane* vapor at the surface at the liquid spill.
b. If a spark is placed just at the surface, will the spill ignite? Explain the reason for your answer.
c. Assume a linear distribution of octane vapor over the boundary layer where it is 4 cm in thickness. Determine the region that the mixture is flammable.

Solution:

1. molecular weight = 12 * 8 + 18 = 114

$$x = e^{-\frac{h_{fg}M}{R}\left(\frac{1}{T}-\frac{1}{T_b}\right)}$$

$$x = e^{-\frac{305\times 114}{8.3144}\left(\frac{1}{313}-\frac{1}{398.6}\right)}$$

$$x = 3.65\,\%$$

2. No, it is above UFL.
3. Assume a linear distribution profile as x = ay + b, and apply boundary conditions: y = 0, x = 3.65; y = 4, x = 0. We have, x = −0.9125y + 3.65, which is translated to y = (−x + 3.65) * 1.096. That means, y = (3.65−3.2) * 1.096 = 0.49 cm is the upper boundary for UFL and y = (3.65−0.95)*1.096 = 2.96 cm is the lower boundary for LFL.

3.4.13 Ambient Conditions for Liquid Evaporation

Problem 3.26 Calculate the range of temperature within which the vapor/air mixture above the liquid surface in a can of *n-hexane* at atmospheric pressure will be flammable.

Solution:
For n-Hexane, the molecular weight is 86 g/mol, Tb = 342 K, $h_{fg} = 0.35\,\mathrm{kJ/g}$, Lower flammability limit is 1.2 %. Upper flammability limit is 7.4 %.

$$0.012 = e^{-\frac{0.35\times 1000\times 86}{8.3144}\left(\frac{1}{T_L}-\frac{1}{342}\right)} \rightarrow T_L = 241\,\mathrm{K} = -31.8\,^{\circ}\mathrm{C}$$

$$0.074 = e^{-\frac{0.35\times 1000\times 86}{8.3144}\left(\frac{1}{T_U}-\frac{1}{342}\right)} \rightarrow T_U = 274.5\,\mathrm{K} = 1.3\,^{\circ}\mathrm{C}$$

Problem 3.27 Calculate the range of ambient pressures within which the vapor/air mixture above the liquid surface in a can of n-Decane (n-$C_{10}H_{22}$) will be flammable at 25 °C.

Solution:

$$\log_{10} p^0 = (-0.2185)\frac{10912}{298} + 8.2481 = 0.2472 \rightarrow p^o = 1.7668 \text{ mmHg}$$

$$x_L = 0.0075 = \frac{1.7668}{p_a} \rightarrow p_a = 235.6 \text{ mmHg}$$

$$x_U = 0.056 = \frac{1.7668}{p_a} \rightarrow p_a = 31.55 \text{ mmHg}$$

Chapter 4
Thermal Balance Method

Over the past 200 years, a fundamental theory based on energy conservation is never fully established to find critical flammability limits. One possible reason is the fact that the thermal properties are temperature dependent and a threshold for ignition is not widely accepted, for this reason an iterative process must be utilized to find the critical flame temperature iteratively. The iterative process prevents a systematic view on all critical phenomena. This inerative process prevents a systematic view on all critical phenomena. Here we will establish the energy conservation at critical limits based on a few assumptions and then analytical expressions can be developed for flammable boundaries. Thus, we can develop more methods out of the theoretical and experimental data to form a complete methodology.

4.1 Thermal Signature

A close check of the temperature dependence for combustion products in the targeted range (300–3000 K) shows most products have a similar shape in enthalpy, since radiation is not significant in this temperature range. It allows us to keep one species (air is preferred) as the baseline, while others have a scaling number in reference to it. Thus, the exact value of critical adiabatic flame temperature is no longer important, since all species have a relatively constant magnitude in reference to the reference species, which is shown in Table 4.1.

From this table, we can see that the quenching potential for most common species are related to the mass ratio and stay roughly constant in the targeted flame temperature range for critical limits (1300–1800 K). Therefore, we can define the quenching potential of a species as

$$Q_i = \frac{E_i}{E_{air}} = \frac{\left(H^0_{AFT} - H^0_{298.15}\right)_i}{\left(H^0_{AFT} - H^0_{298.15}\right)_{air}} \tag{4.1}$$

T. Ma, *Ignitability and Explosibility of Gases and Vapors*,
DOI 10.1007/978-1-4939-2665-7_4

Table 4.1 The quenching potentials (Q_D) of some gaseous agents (diluents)

Temp (K)	AFT (K)	He	Argon	N_2	O_2	CO_2	Air
Q_D	1300	0.655	0.655	0.991	1.046	1.578	1.000
	1600	0.642	0.642	0.992	1.046	1.603	1.000
	1850	0.632	0.632	0.992	1.046	1.615	1.000
$\frac{MW}{MW_{air}}$		0.069	0.624	0.971	1.110	1.526	1.000

Similarly, we can define the heating potential of a fuel as

$$H_i = \frac{\Delta H_{C,i}}{E_{air}} = \frac{\Delta H_{C,i}}{\left(H^0_{AFT} - H^0_{298.15}\right)_{air}} \tag{4.2}$$

where H is the standard enthalpy (kJ/mol) defined by NIST. The advantage of the above treatment is to remove the dependence on variable critical flame temperature, thus the lengthy iterative process can be bypassed.

In order to set up the thermal balance at critical limits, the following assumptions are needed [15]:

- A thermal system is binary and additive
- The flame structure is similar in conditions to be estimated. Therefore, CAFT is constant
- Since CAFT is a constant, a baseline enthalpy of air can be established to scale other terms
- Due to constant CAFT at mixing, the energy terms are additive and can be manipulated between fuels
- Oxygen calorimetry applies at critical conditions. It connects two critical conditions into one thermal system

Assumptions in mind, a generic combustion reaction is expressed as

$$C_aH_bO_c + C_O(O_2 + 3.773\,N_2) + C_dD \rightarrow C_{st}P + C_dD \tag{4.3}$$

$$\begin{aligned} C_O &= a + b/4 - c/2 \\ C_{st} &= 1 + 4.773 \cdot C_O \end{aligned} \tag{4.4}$$

Using the baseline enthalpy of air, all energy terms for the thermal balance can be expressed as a dimensionless term scaled by the enthalpy of air.

At the critical conditions (LFL and UFL), we can establish the energy balances below.

$$x_L \cdot Q_F + (1 - x_L) = x_L \cdot H_F \tag{4.5}$$

$$x_U \cdot Q_F + (1 - x_U) = x_O \cdot (1 - x_U) \cdot H_O \tag{4.6}$$

$$H_F = C_O \cdot H_O \tag{4.7}$$

The first equation states the energy balance at LFL, which is controlled by energy released from fuel. The second equation states the energy balance at UFL, which is limited by energy released from oxygen. The third equation is a constitutive relationship between fuel and oxygen, or oxygen calorimetry. Solving these equations, the thermal signature can be derived.

$$H_O = \frac{x_U - x_L}{C_O \cdot x_U \cdot x_L - x_O \cdot (1 - x_U) \cdot x_L} \tag{4.8}$$

$$Q_F = 1 - 1/x_L + C_O H_O \tag{4.9}$$

From the basic parameters, C_o, x_L and x_U, new intermediate energy terms (Q_F, H_O and H_F) are derived. Since they are unique to that fuel, they can be treated as the thermal signature of that fuel. Reversing back, we have the defining equations of LFL/UFL as

$$x_L = \frac{1}{1 + C_O H_O - Q_F} \tag{4.10}$$

$$x_U = \frac{0.2095 H_O - 1}{0.2095 H_O - 1 + Q_F} \tag{4.11}$$

From these two definitive equations, the lower flammability limit can be interpreted as **the minimum excess heating potential of fuel ($H_F - Q_F$) heating one unit of air ($Q_A = 1$) to CAFT,** while the upper flammability limit is **the minimum excess heating potential of air** (0.2095 $\cdot$ H_O − **1, due to oxygen consumption, since fuel is in excess) heating one unit of fuel** (Q_F) **to CAFT.**

The critical energy balance at flammable limits can be demonstrated graphically. A Heating/Quenching Ratio (HQR), which is defined as the ratio between the heating potential and the quenching potential, can demonstrate the relative competition of energy terms. The cumulative heating potential may be zero, so it can only serve in the **numerator.**

$$HQR_1 = \frac{H_F \cdot x_F}{Q_F \cdot x_F + 1 - x_F} \tag{4.12}$$

$$HQR_2 = \frac{0.2095 \cdot (1 - x_F) \cdot H_O}{Q_F \cdot x_F + 1 - x_F} \tag{4.13}$$

New variables show the heat balance between quenching and heating. If $HQR_1 = 1$, the mixture is at its lower limit. If $HQR_2 = 1$, the mixture is at its upper limit. Flammability limits are defined as $HQR = \min(HQR_1, HQR_2) = 1$. The HQR curves are demonstrated in Fig. 4.1.

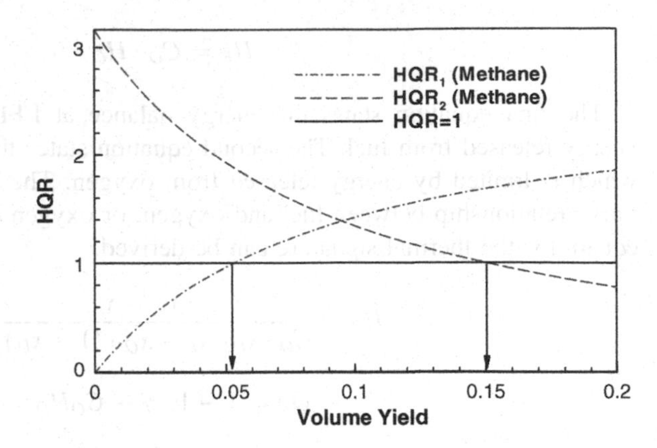

Fig. 4.1 *HQR* curves for determining x_L and x_U of methane

4.2 Correlations for Flammability Limits

Given the expression of LFL and UFL in Eqs. 4.10 and 4.11, we can correlate the excess heating potential term as

$$H_F - Q_F = 21.809 \cdot \Delta H_C \tag{4.14}$$

Or

$$H_F - Q_F = 9.0454 \cdot C_O \tag{4.15}$$

Thus, we have two correlations for LFL.

$$x_L = \frac{1}{1 + 21.809 \cdot \Delta H_C} \tag{4.16}$$

$$x_L = \frac{1}{1 + 9.0454 \cdot C_O} \tag{4.17}$$

Equations 4.16 and 4.17 are checked against experimental data, along with other correlations provided in Sect. 2.2, displayed in Fig. 4.2.

Figure 4.2 shows that correlations for LFL are similar to each other, except those with a polynomial regression. The latter fails if the stoichiometric oxygen number of the fuel is outside of the tested range.

An interesting empirical rule for LFL [43, 64] is that for paraffin hydrocarbons, $x_{st} = 87$ mg/l and $x_L = 48$ mg/l for those with the number of carbon atoms more than 3. Here we can derive the same rules using oxygen calorimetry. The total oxygen consumption for a combustion system is $x_L \cdot C_O$ @ LFL and $x_{st} \cdot C_O$ @

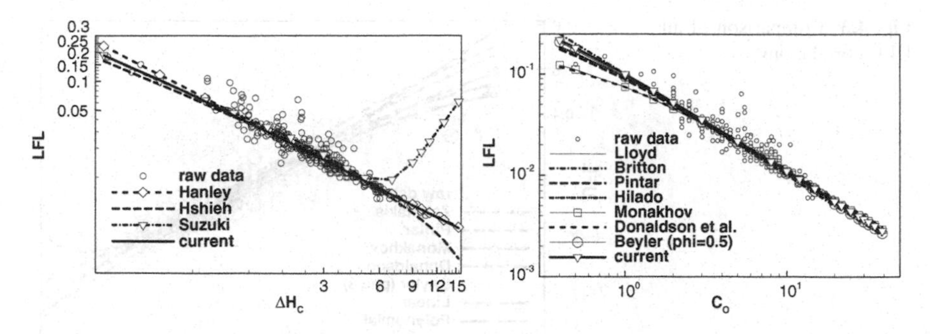

Fig. 4.2 Correlations based on heat of combustion and oxygen calorimetry

stoichiometric point. In addition, the oxygen calorimetry is 414 kJ/mol and a typical heat release rate for paraffin fuel is 44 kJ/g or 0.044 kJ/mg. Therefore, we have

$$x_L \cdot C_O = \frac{C_O}{1 + 9.0454 \cdot C_O} \approx 0.11 \times \frac{414 \text{ kJ/mol}}{22.4 \text{ l/mol}} \times \frac{\text{mg}}{0.044 \text{ kJ}} = 46.2 \text{ mg/l} \quad (4.18)$$

$$x_{st} \cdot C_O = \frac{C_O}{1 + 4.773 \cdot C_O} \approx 0.2095 \times \frac{414 \text{ kJ/mol}}{22.4 \text{ l/mol}} \times \frac{\text{mg}}{0.044 \text{ kJ}} = 88 \text{ mg/l} \quad (4.19)$$

It is not surprising since the correlation in this work is close to Jones' Rule, which was used in deriving these two numbers ($x_{st} = 87$ mg/l and $x_L = 48$ mg/l) [64] (see also Fig. 2.3).

For correlations of UFL, the test data is so scattered that no consistent correlation is widely accepted. Here based on 282 CHON fuels (removing some fuels with a negative H_O and all fuels with F/Cl/Br which are known to change flame temperature in combustion), two correlations are made.

$$x_U = \frac{1}{1 + 2.8648 \cdot \Delta H_C} \quad (4.20)$$

$$x_U = \frac{1}{1 + 1.1843 \cdot C_O} \quad (4.21)$$

They are compared in Fig. 4.3, along with the classical correlations in Sect. 2.2 (see p. 24).

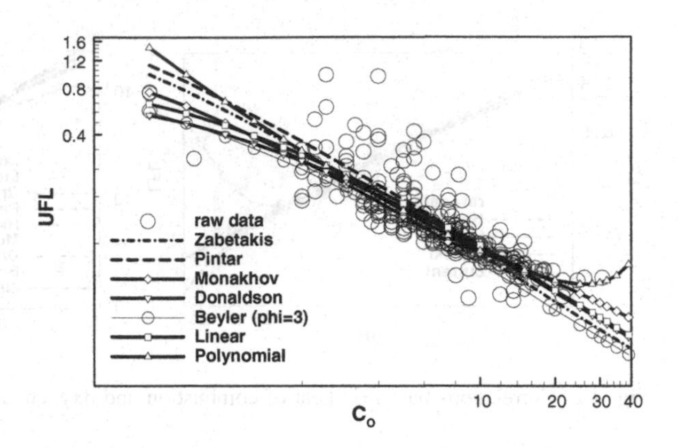

Fig. 4.3 Comparison of all UFL correlations

4.3 The Flammability Limits of a Mixture

4.3.1 Le Chatelier's Rule (LCR)

For the flammability of a flammable fuel mixture, the flammability limits of a mixture are estimated based on Le Chatelier's rule:

$$x_U = \frac{100}{\sum \frac{x_i}{x_{U,i}}} \tag{4.22}$$

$$x_L = \frac{100}{\sum \frac{x_i}{x_{L,i}}} \tag{4.23}$$

However, Le Chatelier's Rule only works with pure fuel species. With diluents involved in a mixture, the fuel-diluent combination will generate a pseudo fuel, with new limits readable from the diluted flammability diagram. Then the Le Chatelier's Rule [68] can be applied to these pseudo fuels.

4.3.2 Thermal Balance Method (TBM)

For flammable mixtures, we have additive relationships like

$$\begin{aligned} &\sum_i x_i = 1 \\ &C_{O,m} = \sum_i C_{O,i} \cdot x_i \\ &Q_{F,m} = \sum_i Q_{F,i} \cdot x_i \\ &H_{F,m} = \sum_i H_{F,i} \cdot x_i \end{aligned} \tag{4.24}$$

Then, we reverse the process to find x_L and x_U for the mixture. Starting with Eq. 4.9, we have

$$1/x_L = 1 + C_O H_O - Q_F \tag{4.25}$$

$$x_i/x_L = x_i + x_i C_O H_O - x_i Q_F \tag{4.26}$$

$$\sum (x_i/x_L) = \sum x_i + \sum (x_i C_O H_O) - \sum (x_i Q_F) = 1 + H_{F,m} - Q_{F,m} \tag{4.27}$$

Therefore, we have

$$x_{L,m} = \frac{1}{1 + H_{F,m} - Q_{F,m}} \tag{4.28}$$

Similarly,

$$1/x_U = 1 + \frac{Q_F}{0.2095 H_O - 1} \tag{4.29}$$

$$x_i/x_{U,i} = x_i + x_i \left(\frac{Q_F}{0.2095 H_O - 1} \right)_i \tag{4.30}$$

$$\sum (x_i/x_U) = \sum x_i + \sum \left(\frac{x_i Q_F}{0.2095 H_O - 1} \right) \tag{4.31}$$

Therefore, we have

$$x_{U,m} = \frac{1}{1 + \sum \frac{Q_{F,i} \cdot x_i}{0.2095 \cdot H_{O,i} - 1}} \tag{4.32}$$

The advantage of this form is that the contribution of a diluent appeared in a form similar to fuel and oxidizer. From the above derivation process, we can also see this method is equivalent to Le Chatelier's Rule (LCR). However, LCR is based on the flammability limits of fuels only, so it will be difficult to incorporate the contribution of a diluent. This method expands the application range of LCR.

4.3.3 Beyler's Method

Another comparable method reported in the literature is Beyler's energy-balance method [79], which is similar to this work in that an energy balance is performed at the Critical Adiabatic Flame Temperature (CAFT). Though integration is used in Eq. 4.33, the averaged specific heat of each product is used to avoid the numerical integration process. The original idea was to determine the flammable mixture of unknown origin with a thermal balance to determine if a reaction would run away (burn). Therefore, it can be used to test the flammability of a mixture.

$$\sum_{i=1}^{n} \frac{\frac{c_i \Delta H_{C,i}}{100}}{\int_{T_0}^{T_f} n_p c_p dT} \geq 1 \tag{4.33}$$

Beyler's method relies on the heat balance of combustion products only because the specific heat of a fuel cannot be retrieved from any reference. The advantage of this treatment is the possibility of incorporating the equivalence ratio or final-state temperature into the estimation scheme, thus allowing the vitiated combustion in a ceiling layer to be addressed.

4.3.4 Estimation Using Correlations

Since the flammability limits of a pure fuel can be estimated from its heat of combustion or the oxygen calorimetry it is natural to estimate the flammability of a mixture based on its mixture energy intensity or mixture stoichiometric oxygen number. Following Hanley's example [47], we can derive the following methodology.

Since the energy terms are additive, it is easy to get mixture energy term ΔH_C or the mixture mass term C_O if the components are already in gaseous state.

$$\Delta H_C^{eff} = \sum_i x_i \cdot \Delta H_{Ci} \tag{4.34}$$

$$C_O = \sum_i x_i \cdot C_{Oi} \tag{4.35}$$

Using correlations in Sect. 4.2, it is possible to estimate flammability limits of a mixture from its constituents.

4.3.5 Non-conventional Estimations

For predicting the flammability limits of complex mixtures there is a family of non-traditional methods called neural network techniques, or artificial intelligence [99]. Using critical variables of structural groups comprising training data from the critical variables of each structural groups, we can accurately predict the flammability limit of an analogous complex mixture having similar structure groups. These critical variables utilize the compositional and thermochemical data from each of the structural groups to produce a neural network model. Through testing the training data from the neural network model, the training and tested data from the neural network can be validated [100]. Since this method is not based on fundamental principles of energy conservation, it will not be explored here.

4.4 Temperature Dependence

By embedding the temperature-dependent variation of the reference species we will find out the temperature dependence of flammability limits. Fortunately, the reference species in the thermal balance method is air, which has a definite thermal property. Equation 4.36 is a simple correlation for characterizing the enthalpy of air, derived from a more-complex correlation provided in NIST chemistry website [101]. Note this correlation is taking T/1000 as input, so the coefficients can have more valid digits. This correlation has been utilized to find the flammability limits of a hydrocarbon mixture [102].

$$E_{air} = \left(H^0_{AFT} - H^0_{298.15}\right)_{air} = f(T) = 1.4893\mathrm{T}^2 + 29.862\mathrm{T} - 9.381 \quad (4.36)$$

Next, define a temperature-dependent enthalpy-scaling factor $\eta(T)$ for measuring the system enthalpy change in reference of air.

$$\eta(T) = \frac{E_i}{E_{air}} = \frac{\left(E^{1600}_{air} - E^{T}_{air}\right)}{\left(E^{1600}_{air} - E^{298}_{air}\right)} = \frac{42.21 - \left(1.4893\mathrm{T}^2 + 29.862\mathrm{T} - 9.381\right)}{42.21 - (-0.35)} = 1.212 - 0.035 \cdot T^2 - 0.702 \cdot T \quad (4.37)$$

Then, apply the energy conservation of flammability limits adjusted by the enthalpy-scaling factor $\eta(T)$.

$$x_L \cdot Q_F \cdot \eta + (1 - x_L) \cdot \eta = x_L \cdot H_F \quad (4.38)$$

$$x_U \cdot Q_F \cdot \eta + (1 - x_U) \cdot \eta = 0.2095 \cdot (1 - x_U) \cdot H_O \quad (4.39)$$

Finally, we have the modified lower flammable limits as

$$x_L = \frac{\eta}{\eta + C_O H_O - Q_F \cdot \eta} \quad (4.40)$$

This expression can be rewritten as

$$\frac{x_L}{x_{L,0}} = \frac{\frac{\eta}{\eta + C_O H_O - Q_F \cdot \eta}}{\frac{1}{1 + C_O H_O - Q_F \cdot 1}} = \eta \cdot \frac{1 + C_O H_O - Q_F \cdot 1}{\eta + C_O H_O - Q_F \cdot \eta} = \eta \cdot \eta_f \quad (4.41)$$

where η is the temperature-variation of air and η_f represents the contribution of added fuel.

$$\eta_f = \frac{1 + C_O H_O - Q_F \cdot 1}{\eta + C_O H_O - Q_F \cdot \eta} \tag{4.42}$$

This temperature dependence has two parts, one (η) is coming from the background species (air), the other (η_f) is a fuel-related correction term, which is usually ignored by all previous work. Since it can be dropped here, a simple linear enthalpy method (designated as LinearH) is proposed to show the difference with and without this term. Therefore, we have a new temperature dependence with a variable enthalpy only.

$$\frac{x_L}{x_{L,0}} = \eta = \frac{E_i}{E_{air}} = \frac{\left(E_{air}^{1600} - E_{air}^{T}\right)}{\left(E_{air}^{1600} - E_{air}^{298}\right)} \tag{4.43}$$

In order to compare the temperature correlations, the experimental data by Kondo et al. [61] are used in Fig. 4.4.

Figure 4.4 shows the temperature dependence of four fuels compared with experimental data. LinearT (Eq. 2.6) is under-predicting the slope, since the mixture properties are not simple functions of temperature only. It oversimplifies the inherent temperature dependence. As expected, LinearH (Eq. 4.43), Zabetakis

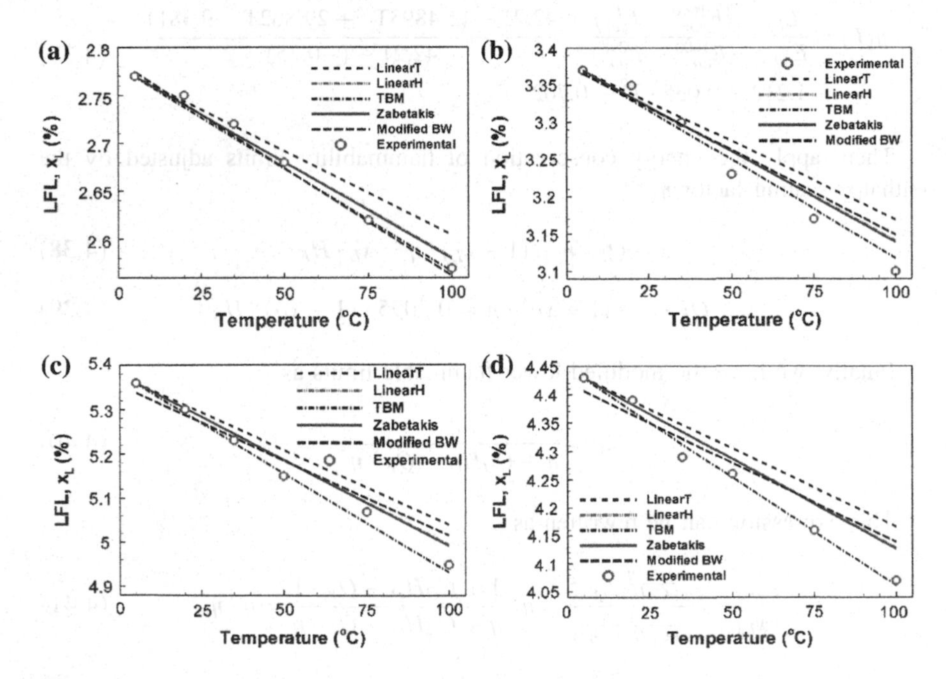

Fig. 4.4 Temperature dependence of lower flammable limits for some fuels. **a** Ethylene. **b** Dimethyl ether. **c** Methyl formate. **d** HFC-152a

(Eq. 2.5) and ModifiedBW (Eq. 2.4) produce similar results, since they are correctly capturing the enthalpy change in the background species air. TBM (Eq. 4.41) produces the best performance, since it takes both the contribution from the temperature-modified air (η) and the contribution of the added fuel (η_f). This also explains the discrepancy in Fig. 2.2.

Most flammability-related theories are built upon the assumption of constant energy absorption of air at ignition, or $x_L \cdot \Delta H_C = K$. It is implicitly assumed that the net energy release from the fuel is absorbed by a constant fraction of air, which assumes the fuel has the same properties of air. So the variation of fuel properties is in fact not considered. White [44] and Zabetakis et al. [49] realized this shortcoming and proposed $x_L \cdot \Delta H_C + \Delta H = K$, where ΔH is the fraction of energy absorbed by the fuel. Based on Eq. 4.41, this method derives $x_L \cdot \Delta H_C = 1117 \cdot (1 - x_L)$ [17] which can be expressed as $\lambda_2 \cdot \frac{\Delta H_C}{C_O} = 1117$. Therefore, this method is assuming a constant energy release from the required oxygen to match the fuel, while all other methods are based on a constant energy release from the fuel. Since flammability is the ability of air to support flame propagation, a constant oxygen-based energy release is more reasonable. Both are derived from the principle of energy conservation and the difference is small (Critical fuel concentration x_L is usually small in a fuel/diluent/air mixture). However, since this method explicitly covers the contribution of fuel, the temperature dependence is better predicted.

4.5 Reconstruction of Flammability Diagrams

The flammability problem of a mixture is difficult since it typically involves three components with dual functions (heating and quenching) in the combustion. Fuel is not only a source of energy, but also a heat absorber during the ignition process. Oxygen is not only a source of energy, but also a heat absorber affecting the flame temperature. Only nitrogen is a typical diluent without any heating role involved. Because of such a ternary nature, the concept of flammability is difficult to present without the help of a flammability diagram. Various diagrams are available to demonstrate the safety operation on a flammable mixture. There is a need to recheck the utility, limitation and theories behind these diagrams.

4.5.1 Standard Flammability Diagrams

One advantage of the Thermal Balance Method is the incorporation of inerting agents (diluents) directly in computing flammability. Now it allows us to have a new look at flammability diagrams. Here, a diluent is introduced into the system with a concentration of x_D and a quenching potential of Q_D. Thus, we have

$$x_L = \frac{1-(1-Q_D)\cdot x_D}{1-Q_F+C_O\cdot H_O} = x_{L,0}\cdot[1-(1-Q_D)\cdot x_D] \tag{4.44}$$

$$x_U = x_{U,0} - \left(x_{U,0} + \frac{Q_D}{Q_F-1+H_O/4.773}\right)\cdot x_D \tag{4.45}$$

Here, the subscript 0 is used to denote the initial flammability limits without any diluent. These two curves produce a flammable envelope (Fig. 4.5). The inerting concentration is the cross-point of the x_L and x_U curves. An additional curve, the stoichiometric line, is also displayed. This is governed by the stoichiometric reaction line (Eq. 4.46).

$$x_{st} = \frac{1-x_D}{1+4.773\cdot C_O} = x_{st,0}(1-x_D) \tag{4.46}$$

It should be noted that the theoretical curves are based on an isothermal process (assuming constant flame temperature at all critical conditions), so there is an inertion point clearly defined. This point usually lies to the right of experimentally determined points [103]. Macek et al. [104] suggested that the flame temperature is raised at extinction (due to the change in flame structure, a higher temperature is needed to sustain the flame), so less agent is required (due to the rise in temperature). As Beyler [64] concludes, the flame temperature is higher at extinction (premixed flame) than at ignition (diffusion flame) due to a distorted flame structure. Shebeco et al. [103] tried to approximate this inertion point more precisely, though, at a price of increased complexity.

When applying this treatment to hexane, the prediction shows significant deviations. The inertion point is on the stoichiometric line of CO (Eq. 4.47) instead of that of CO_2 (Eq. 4.3). This shift is attributed to incomplete combustion and to the preferential diffusion of reactants [64]. Such a shift is clearly shown in the flammability diagrams for propane and hexane in Fig. 4.6. The inertion points appear to be on the stoichiometric line favoring the partial reaction to CO.

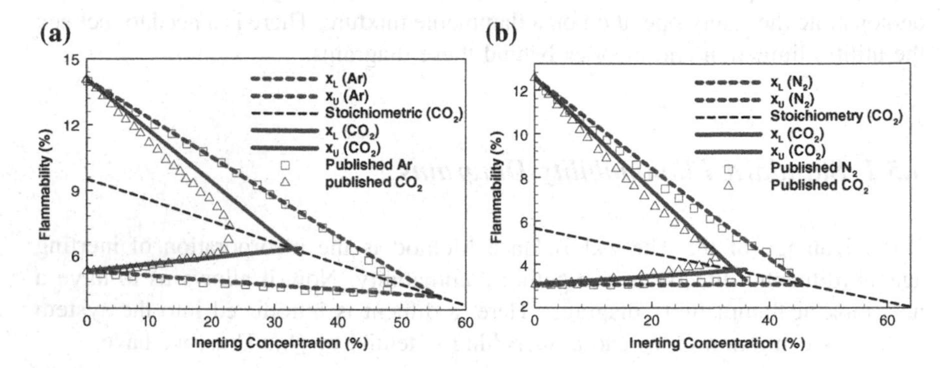

Fig. 4.5 Standard flammability diagrams for two fuels. **a** Methane, **b** ethane

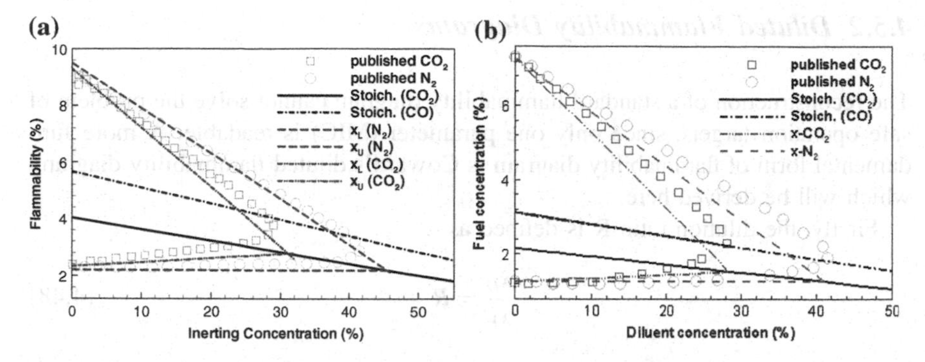

Fig. 4.6 Standard flammability diagrams without and with significant incomplete reaction. **a** propane, **b** hexane

$$
\begin{aligned}
C_O &= a/2 + b/4 - c/2 \\
C_{st} &= 1 + 4.773 \cdot C_O
\end{aligned}
\tag{4.47}
$$

The above-predicted diagrams are based on a constant flame temperature under critical conditions, which lead to a sharp inertion point. The real inertion happens with a higher flame temperature and incomplete reactions, so experimental data will produce a curvature near inertion. To include an incomplete reaction near the inertion point we can project the existing flammable envelope from one thermal agent to another. Figure 4.7 shows the scaling of flammable envelopes by various agents; details can be found in [15].

The rescaled flammability diagram is shown in Fig. 4.7. The data points from one curve (helium) are projected onto other curves (CO_2 and N_2). Such a conversion process retains some features of an incomplete reaction by rescaling the contribution of a diluent only. This idea is useful for some refrigerants, which have an estimated quenching potential by mass scaling.

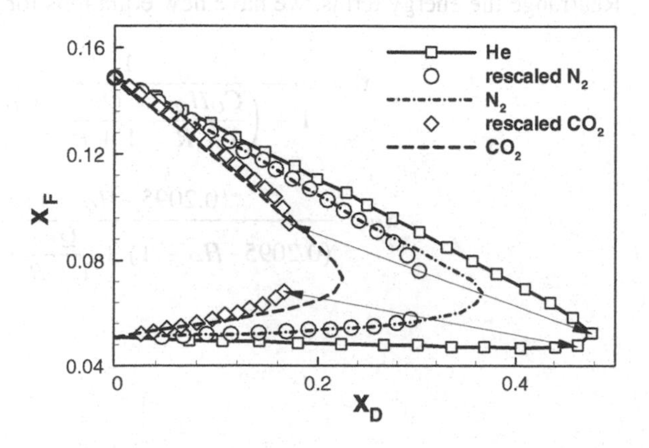

Fig. 4.7 Scaling of flammable envelopes in a standard flammability diagram

4.5.2 Diluted Flammability Diagrams

The reconstruction of a standard flammability diagram cannot solve the problem of safe operation targets, since only one parameter (MIC) is readable. A more fundamental form of flammability diagram is Coward's diluted flammability diagram, which will be derived here.

Firstly, the dilution ratio R is defined as

$$\frac{x_D}{x_F} = R \tag{4.48}$$

Combined with a defining equation $x_F + x_D = x_L$, we have

$$x_D = \frac{R \cdot x_L}{1+R} \tag{4.49}$$

$$x_F = \frac{x_L}{1+R} \tag{4.50}$$

Submit them into the energy conservation equation at LFL

$$x_F \cdot Q_F + x_D \cdot Q_D + 1 - x_L = x_F H_F \tag{4.51}$$

We have the energy equation at UFL

$$\frac{x_L}{1+R} \cdot Q_F + \frac{R \cdot x_L}{1+R} \cdot Q_D + 1 - x_L = \frac{x_L}{1+R} \cdot C_O H_O \tag{4.52}$$

Similarly, we have

$$\frac{x_U}{1+R} Q_F + \frac{R x_U}{1+R} Q_D + 1 - x_U = 0.2095(1 - x_U) H_O \tag{4.53}$$

Rearrange the energy terms, we have new equations for LFL and UFL respectively.

$$x_L = \frac{1}{1 + \left(\frac{C_O H_O}{1+R} - \frac{Q_F}{1+R} - \frac{Q_D \cdot R}{1+R} \right)} \tag{4.54}$$

$$x_U = \frac{(0.2095 \cdot H_O - 1)}{(0.2095 \cdot H_O - 1) + \frac{Q_F}{1+R} + \frac{Q_D \cdot R}{1+R}} \tag{4.55}$$

Forcing $x_L = x_U$, we have the

$$R_{LU} = \frac{0.2095 C_O H_O - 0.2095 \cdot Q_F - C_O}{0.2095 \cdot Q_D} \tag{4.56}$$

A series of flammability diagrams based on Eqs. 4.54 and 4.55 are listed in Fig. 4.8.

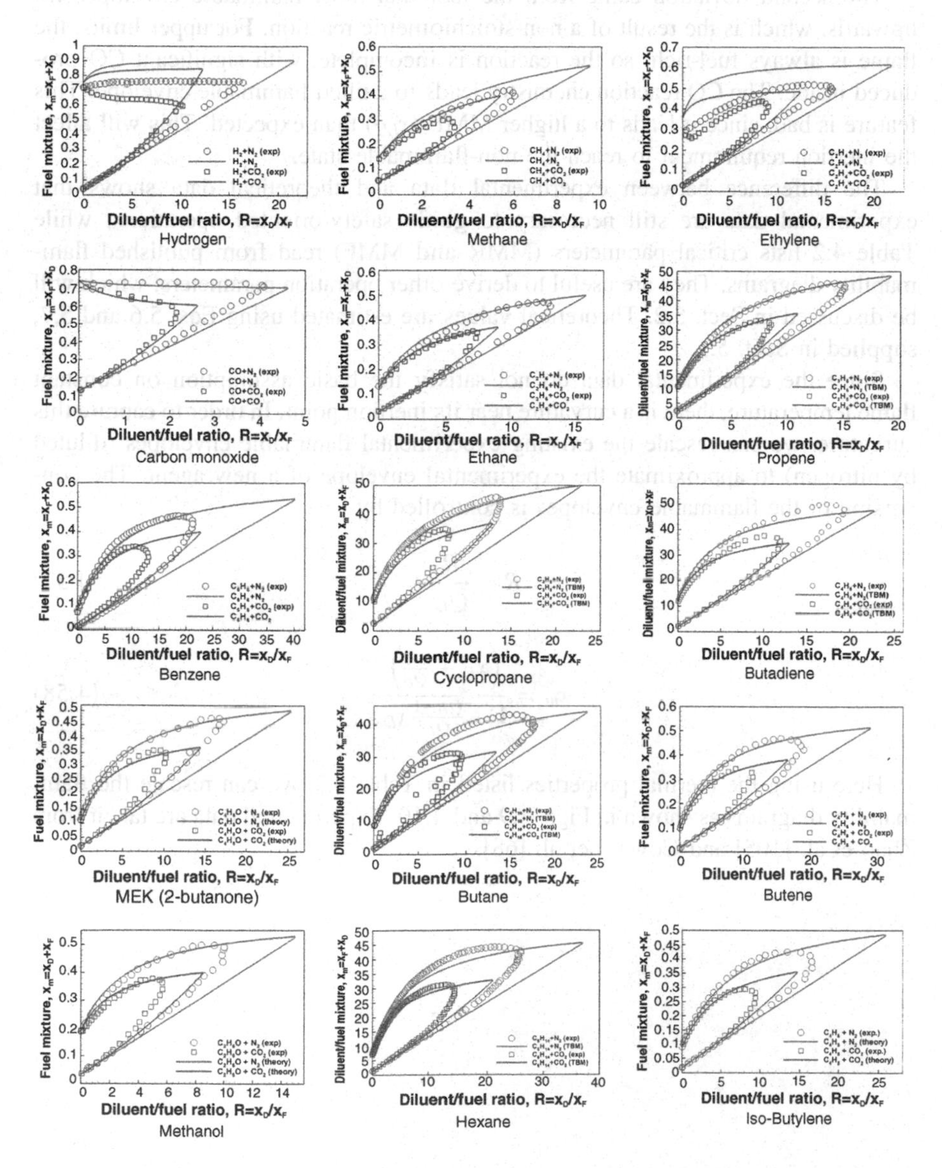

Fig. 4.8 Theoretical diluted flammability diagrams for common fuels (Data taken from [4, 43])

Figure 4.8 shows a series of common fuels with experimental and theoretical flammability diagrams. There are two major differences from experimental envelopes to theoretical envelopes. First, since extinction occurs at a higher flame temperature, the assumption on constant flame temperature breaks down near the inertion point, so the experimental envelopes are always shorter than theory. This is a good feature, which means the dilution requirement based on a theoretical inertion point is generally more conservative than using experimental data.

The second deviation came from the fact that most flammable envelopes tilt upwards, which is the result of a non-stoichiometric reaction. For upper limits, the flame is always fuel-rich, so the reaction is incomplete, with significant CO produced inside. The CO reaction chemistry leads to a tilted flammable envelope. This feature is bad, since it leads to a higher MMF (x_{LU}) than expected. This will affect the inertion requirement to reach the non-flammable state.

The difference between experimental data and theoretical data shows that experimental data are still necessary to guide safety-oriented operations, while Table 4.2 lists critical parameters (MMR and MMF) read from published flammability diagrams. They are useful to derive other operation parameters, which will be discussed in Sect. 5.2. Theoretical values are estimated using Eqs. 5.6 and 5.7, supplied in Sect. 5.1.

Since the experimental data do not satisfy the basic assumption on constant flame temperature, there is a curvature near its inertion point. In order to capture this curvature, we can rescale the existing experimental flammable envelopes (diluted by nitrogen) to approximate the experimental envelope of a new agent. The conversion of the flammable envelopes is controlled by

$$R = \frac{R}{Q_D} \tag{4.57}$$

$$x_m = \frac{\left(x_F + \frac{x_D}{Q_D}\right)}{1 - \frac{Q_D - 1}{Q_D} x_D} \tag{4.58}$$

Here using the thermal properties listed in Table 4.3, we can rescale the flammability diagrams as shown in Figs. 4.9 and 4.10. Experimental data are taken from Zhao et al. [105] and Coward et al. [68].

Table 4.2 Critical parameters for a fuel, taken from the diagrams in [4, 43]

Fuel	Formula	C_O	x_L	x_U	exp. R_{LU}	exp. x_{LU}	theo. R_{LU}	theo. x_{LU}
Acetone	C_3H_6O	4.000	0.026	0.128	11.954	0.454	18.517	0.505
Acetylene	–	–	–	–	25.873	0.800	–	–
Avgas	–	–	–	–	24.101	0.434	–	–
Benzene	C_6H_6	7.500	0.013	0.079	21.180	0.430	40.447	0.537
Butadiene	C_4H_6	5.500	0.020	0.120	17.370	0.500	22.930	0.477
Butane	C_4H_{10}	6.500	0.018	0.084	17.550	0.430	23.719	0.443
Butene	C_4H_8	6.000	0.017	0.097	19.001	0.469	29.419	0.515
Carbon disulfide	CS_2	3.000	0.013	0.500	39.921	0.763	62.100	0.815
Carbon monoxide	CO	0.500	0.125	0.740	4.080	0.730	4.651	0.703
Cyclopropane	C_3H_8	5.000	0.015	0.094	13.750	0.460	42.137	0.644
Diethyl ether	$C_4H_{10}O$	6.000	0.019	0.360	26.878	0.517	23.177	0.458
Ethane	C_2H_6	3.500	0.030	0.124	12.880	0.470	15.753	0.501
Ethyl alcohol	C_2H_6O	3.000	0.033	0.190	10.012	0.497	15.104	0.529
Ethylene	C_2H_4	3.000	0.027	0.360	13.160	0.520	21.892	0.615
Ethylene dichloride	$C_2H_4Cl_2$	2.500	0.045	0.173	2.465	0.301	9.364	0.465
F-12	CCl_2F_2	–	–	–	4.159	0.254	–	–
Gasoline	–	–	–	–	24.270	0.446	–	–
Hexane	C_6H_{14}	9.500	0.012	0.074	25.930	0.440	37.286	0.458
Hydrogen	H_2	0.500	0.040	0.750	16.420	0.760	21.788	0.905
Isobutyl formate	$C_5H_{10}O_2$	6.500	0.017	0.080	14.218	0.426	27.013	0.474
Isobutylene	C_4H_8	5.500	0.016	0.100	16.592	0.425	35.531	0.582
JP4	–	–	–	–	26.318	0.448	–	–
MEK	$C_6H_{12}O$	8.500	0.012	0.080	17.253	0.474	42.097	0.515
Methane	CH_4	2.000	0.050	0.150	6.005	0.423	9.530	0.524
Methyl acetate	$C_3H_6O_2$	3.500	0.032	0.160	10.338	0.484	13.653	0.467
Methyl alcohol	CH_4O	1.500	0.067	0.360	5.691	0.540	6.820	0.522
Methyl butene	C_5H_{10}	7.500	0.015	0.091	22.604	0.462	30.108	0.465
Paraffin	–	–	–	–	54.263	0.425	–	–
Pentane	C_5H_{12}	8.000	0.014	0.078	23.483	0.455	32.502	0.467
Propylene	C_3H_6	4.500	0.024	0.110	15.140	0.460	19.342	0.486

Table 4.3 Thermal properties for major refrigerants

Refrigerant	Formula	MW (g/mol)	C_O	x_L	x_U	Q_F or Q_D	H_O
R32	CH_2F_2	52.02	1.0	0.127	0.335	1.5123	8.3646
R143a	CF_3CH_3	84.04	2.0	0.07	0.19	3.0539	8.1459
R152a	CHF_2CH_3	66.05	2.5	0.042	0.202	5.4215	11.2721
R290	C_3H_8	44.10	5.0	0.022	0.095	13.7849	11.6198
R600	C_4H_{10}	58.02	6.5	0.018	0.084	12.7209	10.3502
R134a	CF_3CH_2F	102.03				3.537	
R227ea	CF_3CHFCF_3	170.03				5.896	

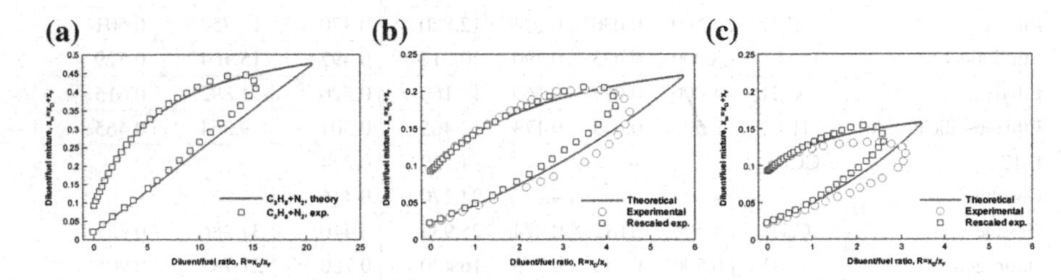

Fig. 4.9 Flammability diagrams for R290 (propane). **a** Diluted by nitrogen, **b** diluted by R134a **c** diluted by R227ea

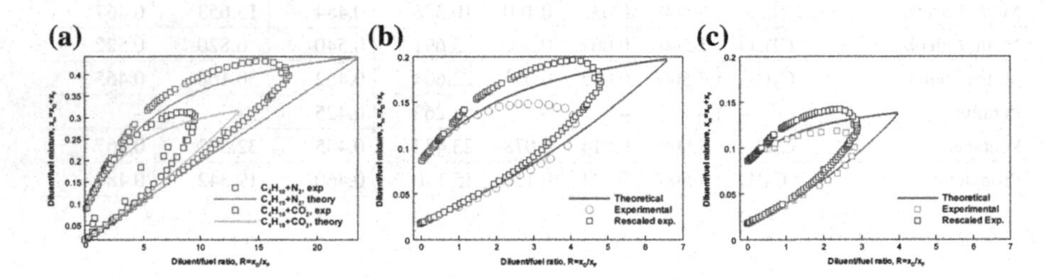

Fig. 4.10 Flammability diagrams for R600 (n-butane). **a** Diluted by nitrogen, **b** diluted by R134a **c** diluted by R227ea

4.6 Problems and Solutions

4.6.1 Temperature Dependence

Problem 4.1 The lower flammable limit of propane at 20 °C is 2.1 % by volume. Find the lower flammable limit at 200 °C.

Solution:

1. Zabetakis' method:

$$\frac{x_L}{x_{L,0}} = 1 - \frac{0.75}{x_{L,0} \cdot \Delta H_C} \cdot (T - T_0) = 1 - 0.000721 \cdot (T - T_0) = 1 - 0.00721 \times (200 - 20) = 0.87022$$

$$x_L = 1.83$$

2. Beyler's method:

$$\frac{x_L}{100} \Delta H_C = nC_p \left(T_{f,L} - T_0\right)$$

Since $T_{f,L} = 1600\,K$, $T_0 = 293\,K$, $x_L = 2.1$, we have

$$\frac{\Delta H_C}{nC_p} = \frac{\left(T_{f,L} - T_0\right)}{x_L/100} = \frac{1600 - 293}{2.1/100} = 6.22 \times 10^4\,K$$

At the new ambient temperature, $\frac{\Delta H_C}{nC_p} = \frac{\left(T_{f,L} - T_0\right)}{x_L/100} = \frac{1600 - 493}{x_L/100} = 6.22 \times 10^4\,K$, $x_L = 1.8$

3. Thermal balance method:

$$\eta(200) = 1.212 - 0.035 \times 0.2^2 - 0.702 \times 0.2 = 0.872$$

$$\frac{x_L}{x_{L,0}} = \frac{\frac{\eta}{\eta + C_O H_O - Q_F \cdot \eta}}{\frac{1}{1 + C_O H_O - Q_F \cdot 1}} = \eta \cdot \frac{1 + C_O H_O - Q_F \cdot 1}{\eta + C_O H_O - Q_F \cdot \eta} = 0.872 \times \frac{1 + 5 \times 12.34 - 15.23}{0.872 + 5 \times 12.34 - 15.23 \times 0.872} = 0.840$$

$$x_L = 1.76$$

4.6.2 Flammable State with Multiple Fuels

Problem 4.2 A group of Japanese researchers tested the following fuel combinations for mixture flammability [106]. You are required to

a. Find the thermal signature of each fuel and fill into blanks;

	Ethylene (C_2H_4)	Dimethyl ether (C_2H_6O)	Carbon monoxide (CO)
LFL	0.027	0.033	0.122
UFL	0.315	0.262	0.725
Hc	1323.1	1328.4	283.0
C_O	**3.0**	**3.0**	**0.5**
H_O	**13.314**	**11.0**	**16.219**
Q_F	**3.905**	**3.697**	**0.912**

b. Estimate the lower flammable limit (LFL) for each combination of fuels.

No.	C_3H_8	C_2H_4	C_2H_6O	CO	Observed (vol%)	LCR	TBM	OBC	EBC
86	0	0.33	0.33	0.34	4.02	4.00	4.00	4.51	4.89

Solution:

1. Find the thermal signature using Eqs. 4.8 and 4.9.
2. Sum up the energy terms for a mixture

$$C_{O,m} = \sum_i C_{O,i} \cdot x_i = 0.33 \times 3 + 0.33 \times 3 + 0.34 \times .5 = 2.15$$

$$\Delta H_{C,m} = \sum_i \Delta H_{C,i} \cdot x_i = 0.33 \times 1.3231 + 0.33 \times 1.3284 + 0.34 \times 0.283 = 0.9712 \text{ MJ/mol}$$

$$H_{O,m} = \sum_i H_{O,i} \cdot x_i = 0.33 \times 13.314 + 0.33 \times 11.0 + 0.34 \times 16.219 = 13.538$$

$$Q_{F,m} = \sum_i Q_{F,i} \cdot x_i = 0.33 \times 3.905 + 0.33 \times 3.697 + 0.34 \times 0.912 = 2.819$$

3. Reverse back into the concentrations

$$\text{OBC}: x_L = \frac{1}{1 + 9.0454 \cdot C_O} = \frac{1}{1 + 9.0454 \times 2.15} = 0.0489$$

$$\text{EBC}: x_L = \frac{1}{1 + 21.809 \cdot \Delta H_C} = \frac{1}{1 + 21.809 \times 0.9712} = 0.0451$$

$$\text{TBM}: x_{L,m} = \frac{1}{1 + (3 \times 13.314 - 3.905) \times .33 + (3 \times 11 - 3.697) \times .33 + (0.5 \times 16.219 - 0.912) \times .34} = 0.0400$$

$$\text{LCR}: x_L = \frac{100}{\sum \frac{x_i}{x_{L,i}}} = \frac{100}{\frac{33}{2.7} + \frac{33}{3.3} + \frac{34}{12.2}} = 4.00\ \%$$

Problem 4.3 A mixture of gases is composed of Ethyl Acetate 63.5 %, Ethanol 20.8 %, and Toluene 15.7 %. Find the flammability limits for such a fuel/diluent mixture [66].

Solution: a typical procedure for thermal balance method involves three steps.

1. Collect inputs of C_O, x_L, and x_U, together with the composition x_i into a spreadsheet. The thermal signature of each fuel is derived using equations.

$$H_O = \frac{x_U - x_L}{C_O \cdot x_U \cdot x_L - 0.2095 \cdot (1 - x_U) \cdot x_L}, \quad Q_F = 1 - 1/x_L + C_O H_O,$$

$$H_F = C_O \cdot H_O$$

2. Sum-up the thermal signature of the mixture.

$$\sum_i y_i = 1; C_{O,m} = \sum_i C_{O,i} \cdot y_i; Q_{F,m} = \sum_i Q_{FA,i} \cdot y_i; H_{F,m} = \sum_i H_{F,i} \cdot y_i; H_{O,m} = H_{F,m}/C_{O,m}$$

Table 4.4 Spreadsheet calculation of a pseudo-fuel with multiple fuels

Fuel	C_O	x_i	x_L	x_U	Q_F	H_O
Ethyl acetate	5.00	0.635	0.0250	0.114	7.370	9.274
Ethanol	3.00	0.208	0.0430	0.190	3.394	8.550
Toluene	9.00	0.157	0.0243	0.071	17.346	11.101
Mixture	5.212	1.000	0.0243	0.110	7.424	9.122

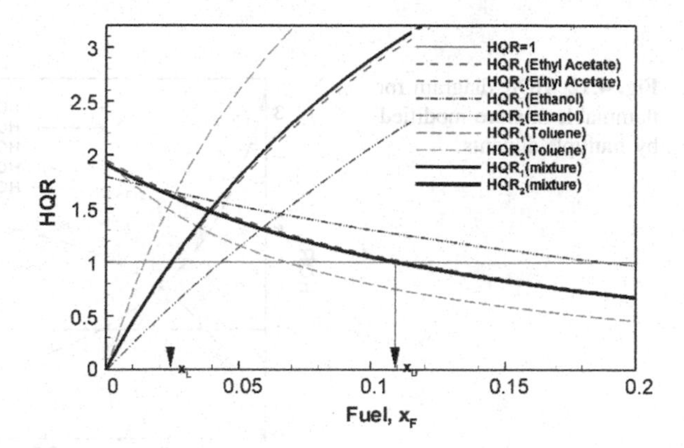

Fig. 4.11 HQR curves modified by multiple fuels

3. Limits of the mixture are estimated using $x_{L,m} = \frac{1}{1+H_{F,m}-Q_{F,m}}$, $x_{U,m} = \frac{1-0.2095 \cdot H_{O,m}}{1-0.2095 \cdot H_{O,m}-Q_{F,m}}$.

A spreadsheet calculation is shown in Table 4.4, whereas the results are displayed in Fig. 4.11.

4.6.3 Flammable State with Multiple Diluents

Problem 4.5 A mixture of gases is composed of Methane 15.2 %, Nitrogen 79.8 %, Carbon Dioxide 0.5 %. Find the flammability limits for such a fuel/diluent mixture.
Solution: Similarly following previous steps.

1. Collect inputs of C_O, x_L, and x_U, together with the composition x_i into a spreadsheet.
2. Sum-up the thermal signature of the mixture.

$$\sum_i y_i = 1; C_{O,m} = \sum_i C_{O,i} \cdot y_i; Q_{F,m} = \sum_i Q_{FA,i} \cdot y_i; H_{F,m} = \sum_i H_{F,i} \cdot y_i; H_{O,m} = H_{F,m}/C_{O,m}$$

Table 4.5 Spreadsheet calculation of a pseudo-fuel with multiple diluents

Input							Intermediate	
Fuel	C_O	x_i	x_L	x_U	Q_D	Q_F	H_O	H_F
CH_4	2.000	0.152	0.050	0.150	–	13.81	16.4	32.8
N_2	–	0.798	–	–	0.992	0.992	0.000	0.000
CO_2	–	0.050	–	–	1.75	1.75	0.000	0.000
Mixture	0.304	1.000	**0.0324**	**0.450**	–	2.98	16.4	32.8

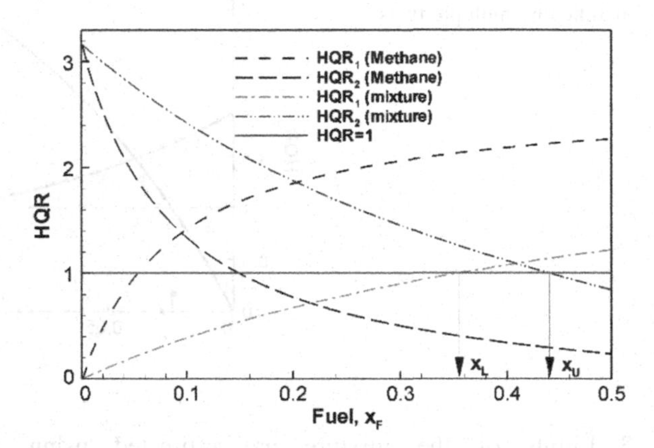

Fig. 4.12 HQR diagram for flammable mixture modified by multiple diluents

3. Limits of the mixture are estimated using $x_{L,m} = \frac{1}{1+H_{F,m}-Q_{F,m}}$, $x_{U,m} = \frac{1-0.2095 \cdot H_{O,m}}{1-0.2095 \cdot H_{O,m}-Q_{F,m}}$.

A spreadsheet calculation is shown in Table 4.5, whereas the results are displayed in Fig. 4.12.

4.6.4 Flammable State of a Coal-Mine-Gas Mixture

Problem 4.6 A sample taken from a sealed area yields the mixture composition as follows: CH_4 ~ 10 %, CO ~ 5 %, N_2 ~ 75 %, O_2 ~ 10 %. Is this gas-mixture flammable?

Solution: This problem can be solved in five approaches.

A. Two fuels plus excess nitrogen burning in air;
B. Two fuels burning in air with excess nitrogen as diluent;
C. Two fuels burning in oxygen-modified air.
D. Using Le Chatelier's Rule
E. Using HQR diagram

Table 4.6 Thermal balance method for flammability range of a fuel/diluent mixture

	C_O	x_L (%)	x_U (%)	Q_F	H_O	H_F	Composition (%)
Methane	2.00	5.00	15.00	13.81	16.38	32.77	19.13
Carbon monoxide	0.50	12.50	74.00	0.80	15.59	7.80	9.57
Nitrogen	–	–	–	0.99	0.00	0.00	71.30
Mixture	0.43	**21.79**	**41.34**	3.43	4.63	7.01	100.00

Approach A: *Excess Nitrogen as a Diluent*

This is a dilution problem where a fuel/diluent mixture (10 % CH_4, 5 % CO and 37.27 % N_2) is immersed in 47.73 % of air (10 % O_2 + 37.73 % N_2).

1. Get the thermal signature from flammability limits of each component (see Table 4.6).
2. Sum up each individual thermal signature into a lump-sum signature for the mixture.
3. Get the flammability limits from the mixture signature.

Conclusion: Current fuel mixture has a concentration of (1 − 47.73 %) = 52.27 %, well above the flammability range of 21.79–41.34 %. So this sample is too rich to ignite.

Approach B: *Excess Nitrogen as Part of a Pseudo Fuel*

This pseudo-fuel is composed of two fuels and diluted by a third agent (excess nitrogen).

a. Get the thermal signature of the pseudo-fuel first (Table 4.7) without dilution.
b. Then a dilution diagram (Fig. 4.13) is reconstructed from this mixture signature.
c. The same flammability range will be reached by either reading the diagram or solving equations using a given R = 37.27 %/15 % = 2.48 (Table 4.7).

The above two approaches are consistently represented in Fig. 4.13. Since the sample point lies above the flammable zone, this sealed area is too rich to ignite.

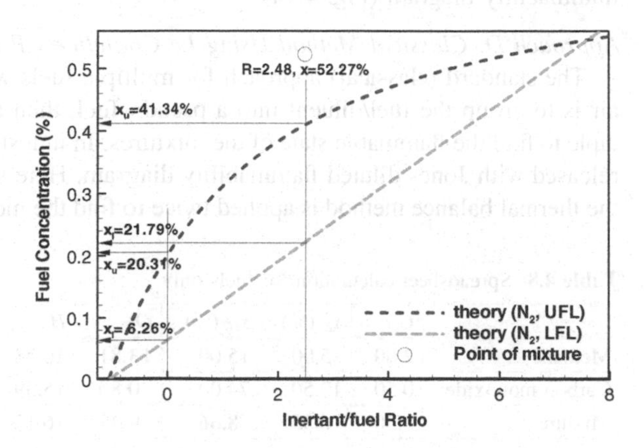

Fig. 4.13 A diluted flammability diagram presenting the flammable state

Table 4.7 Thermal balance method for the thermal signature of a fuel-only mixture

	C_O	x_L (%)	x_U (%)	Q_F	H_O	H_F	Composition (%)
Methane	2.00	5.00	15.00	13.81	16.38	32.77	66.67
Carbon monoxide	0.50	12.50	74.00	0.80	15.59	7.80	33.33
Mixture	1.50	**6.26**	**20.31**	9.47	16.12	24.44	–

Conclusion: Since the operation point lies outside the flammable zone, this mixture is not flammable.

Approach C: Excess Nitrogen as Part of Oxygen-Modified Air

This pseudo-fuel is composed of two fuels and burning in oxygen-modified air.

a. Get the thermal signature of the pseudo-fuel first without dilution (see Table 4.8).
b. Using oxygen fraction to get the flammable limits
c. An oxygen-modified flammability diagram is reconstructed from the thermal signature in Table 4.8.

Here the oxygen fraction is $\lambda = \frac{0.1}{0.1+0.75} = 0.118$, while $Q_m = \lambda Q_O + (1-\lambda)Q_N = 0.118 \times 1.046 + 0.882 \times 0.992 = 0.998$.

Using oxygen-modified flammability equation with R = 0 (no diluent in this case), the flammable range is estimated as

$$x_L = \frac{Q_m}{Q_m + \frac{C_O H_O}{1+R} - \frac{Q_F}{1+R} - \frac{Q_D R}{1+R}} = \frac{0.998}{0.998 + 1.5 \times 16.12 - 9.47} = 0.0635$$

$$x_U = \frac{\lambda \cdot H_O - Q_m}{\lambda \cdot H_O - Q_m + \frac{Q_F}{1+R} + \frac{Q_D \cdot R}{1+R}} = \frac{0.118 \times 16.12 - 0.998}{0.118 \times 16.12 - 0.998 + 9.47} = 0.0866$$

Current measurement of the fuel is 15 %, which is above the flammable range of 6.35–8.66 %, so this mixture is too rich to burn.

The flammable state of this mixture is shown in the following oxygen-modified flammability diagram (Fig. 4.14).

Approach D: Classical Method Using Le Chatelier's Principle

The standard (classical) approach for multiple fuels with at least one diluent in air is to group the fuel/diluent into a pseudo-fuel, then apply Le Chatelier's principle to find the flammable state of the mixtures. In industry, the grouping process is released with Jones diluted flammability diagram. Here without experimental data, the thermal balance method is applied twice to find the modified flammable range of

Table 4.8 Spreadsheet calculation for fuels only

	C_O	x_L (%)	x_U (%)	Q_F	H_O	H_F	Composition (%)
Methane	2.00	5.00	15.00	13.81	16.38	32.77	66.67
Carbon monoxide	0.50	12.50	74.00	0.80	15.59	7.80	33.33
Mixture	1.50	**6.35**	**8.66**	9.47	16.12	24.44	–

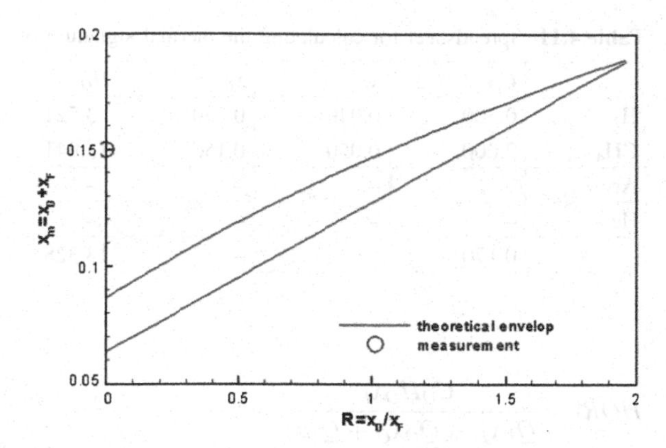

Fig. 4.14 Flammable state in an oxygen-modified flammability envelope

Table 4.9 Grouping 10 % methane with some (30 %) excess nitrogen

	C_O	x_L (%)	x_U (%)	Q_F	H_O	H_F	Composition (%)
Methane	2.00	**5.00**	**15.00**	13.81	16.40	32.81	25.00
Nitrogen	–	–	–	0.99	–	–	75.00
	0.50	**19.98**	**36.73**	4.195	16.40	8.20	100.00

two pseudo fuels. First, 10 % methane is grouped with 30 % excess nitrogen to form a diluted methane, with a flammable range of 10–24.77 % (Table 4.9). Next, 5 % carbon monoxide is grouped with 8.4 % excess nitrogen to form a diluted carbon monoxide, with a flammable range of 38.67–70.93 % (Table 4.10).

Here is the Le Chatelier's rule.

$\frac{0.40}{0.1998} + \frac{0.1227}{0.3065} = 2.40 > 1 \rightarrow$ This means the mixture is above its lower flammable limit.

$\frac{0.4}{0.3673} + \frac{0.1227}{0.7126} = 1.26 > 1 \rightarrow$ This means the mixture is above its upper flammable limit.

Based on these two criteria, the mixture is too rich to ignite.

Approach E: Using Heating-Quenching Ratio

The simplest way to find the mixture flammable state is to compute and compare heating/quenching potential ratio directly. Using values in Table 4.6, we have

Table 4.10 Grouping 5 % CO with the rest (7.27 %) excess nitrogen

	C_O	x_L (%)	x_U (%)	Q_F	H_O	H_F	Composition (%)
CO	0.50	**12.50**	**74.00**	0.80	15.59	7.80	40.75
Nitrogen	–	–	–	0.99	–	–	59.25
	0.204	**30.65**	**71.26**	0.914	15.59	3.18	100.00

Table 4.11 Spreadsheet for calculating the thermal signature of a flammable mixture

	C_O	x_L	x_U	H_O	Q_F	x_i
H_2	0.500	0.040	0.750	3.521	55.010	0.020
CH_4	2.000	0.050	0.150	13.941	16.383	0.080
Ar	–	–	–	–	0.632	0.250
He	–	–	–	–	0.632	0.650
	0.170	–	–	13.328	2.980	–

$$HQR_1 = \frac{C_O H_O x_F}{Q_F x_F + Q_O x_O + Q_D x_D}$$

$$= \frac{32.81 \times 0.1 + 7.8 \times .05}{13.81 \times 0.10 + 0.8 \times 0.05 + 0.992 \times 0.3727 + 1 \times 0.4773} = 3.164 > 1$$

This means the fuel-based energy release is sufficient to bring about ignition, or the mixture is ignitable.

$$HQR_2 = \frac{H_O \cdot x_O}{Q_F x_F + Q_O x_O + Q_D x_D}$$

$$= \frac{16.12 \times 0.15}{13.81 \times 0.1 + 0.8 \times 0.05 + 0.992 \times 0.3727 + 1 \times 0.4773} = 0.306 < 1$$

This means oxygen-based energy release is insufficient to bring about explosive burning, or the mixture is non-explosive.

Based on the above criteria, the mixture is ignitable and non-explosive, above its flammable zone.

Problem 4.7 Let us redo the ISO10156 problem in Sect. 2.3. A mixture has 2 % Hydrogen, 8 % methane, 65 % Helium, and 25 % Argon, is it flammable in air? [71]

Solution:

This is a fuel mixture (two fuels) inerted by two diluents. We can sum-up its thermal signature from each individual constituent (Table 4.11).

$$x_L = \frac{1}{1 + C_O H_O - Q_F} = \frac{1}{1 + .17 \times 13.328 - 2.98} = 3.5 > 1$$

Since the LFL of this mixture is larger than one, so this mixture is non-ignitable in air.

Chapter 5
Ignitability, Flammability and Explosibility

5.1 Two Types of Flammability Problems

Generally, there are two types of problems dealing with mixture safety. **Type I problem** deals with a static mixture in a confined space. It answers the question, **is the mixture self-flammable?** Change of one component (such as fuel) will affect other components (such as oxygen and diluents) accordingly. Therefore, a type I problem always comes with a variable oxygen level. The dynamic process for updating all concentrations in a confined space is called **Purge**, such as the inertion of a confined space with a suppressant. A compartment fire is a typical type I problem, where the oxygen concentration is fluctuating over time. Similarly, the counter-solution to a compartment fire is the total flooding by gaseous agents, which dilutes both oxygen and fuel.

Type II problem deals with a dynamic fuel stream, usually composed of a fuel and a diluent, releasing into air. It answers the question, **is the mixture ignitable in air?** Change of one component (such as the fuel) will affect the accompanying constituent, but not the background air (or oxidizer). Therefore, a type II problem comes with a dynamic fuel stream releasing into a constant background. The dynamic process to change the fuel/diluent ratio is called **Dilution**, while the background (or oxidizer) condition stays the same. A typical type II problem is the dilution of a fuel leak, where the fuel concentration is so diluted that an ignition is not possible in its fully diluted state. In suppression engineering, the local application of a suppressant is to dilute the fuel, so it is also a type II problem.

Since the two types of flammability problems have different applications, they have different representative curves and critical points in flammability diagrams, through which the concepts of ignitability, explosibility and flammability will be introduced along with mathematical expressions.

T. Ma, *Ignitability and Explosibility of Gases and Vapors*,
DOI 10.1007/978-1-4939-2665-7_5

5.1.1 Type I Problem (Variable Oxygen in Background Air)

As part of a combustion reaction, oxygen in the mixture serves as a major controlling factor for supporting flame propagation. Here for an oxygen-modified air with a diluent (nitrogen), the quenching potential of this mixture is defined by

$$Q_m = (1 - \lambda) \cdot Q_D + \lambda \cdot Q_O \tag{5.1}$$

where λ is the molar fraction of oxygen in this nitrogen/oxygen mixture. Applying $\lambda = 0.2095$ for normal air, we have $Q_m = 0.992 \times (1 - 0.2095) + 1.046 \times 0.2095 = 1$, which is the quenching potential of air used as the reference.

Following the thermal balance at critical points, we have the governing equation for LFL defined as

$$\frac{x_L}{1+R} \cdot Q_F + \frac{R \cdot x_L}{1+R} \cdot Q_D + (1 - x_L) \cdot Q_m = \frac{x_L}{1+R} \cdot C_O H_O \tag{5.2}$$

Similarly the governing equation for UFL is

$$\frac{x_U}{1+R} Q_F + \frac{R \cdot x_U}{1+R} Q_D + (1 - x_U) \cdot Q_m = \lambda \cdot (1 - x_U) \cdot H_O \tag{5.3}$$

Solve the above two equations, we have

$$x_L = \frac{Q_m}{Q_m + \frac{C_O H_O}{1+R} - \frac{Q_F}{1+R} - \frac{Q_D R}{1+R}} \tag{5.4}$$

$$x_U = \frac{\lambda \cdot H_O - Q_m}{\lambda \cdot H_O - Q_m + \frac{Q_F}{1+R} + \frac{Q_D R}{1+R}} \tag{5.5}$$

which form the boundary of a flammable envelope.

Forcing $x_L = x_U$, we have the cross point (or the inertion point) of LFL/UFL lines controlled by

$$R_{LU} = \frac{C_O H_O \lambda - \lambda Q_F - C_O Q_m}{\lambda Q_D} \tag{5.6}$$

$$x_{LU} = \frac{Q_m}{Q_m + \frac{C_O H_O}{1+R_{LU}} - \frac{Q_F}{1+R_{LU}} - \frac{Q_D R_{LU}}{1+R_{LU}}} \tag{5.7}$$

European definition of limiting oxygen concentration is to support ignition initiation, or the flammable envelope is non-zero to allow ignition to start [11]. The mathematical realization of this requirement is $R_{LU} = 0$, or $C_O H_O \lambda - \lambda Q_F - C_O Q_m = 0$, which translates to

$$\lambda_2 = \frac{C_O Q_N}{C_O H_O - Q_F - (Q_O - Q_N) \cdot C_O} \tag{5.8}$$

For fuels burning in normal ambient air, since $Q_N = 0.992 \approx 1$, $Q_O - Q_N = 1.046 - 0.992 = 0.054 \approx 0$ and $x_L = \frac{1}{1+C_O H_O - Q_F}$, the limiting oxygen concentration (LOC) is simplified as

$$\lambda_2 \approx \frac{C_O}{C_O H_O - Q_F} = \frac{C_O x_L}{1 - x_L} \tag{5.9}$$

Note λ_2 makes the background air non-explosive, no matter what fuel is used.

American definition of limiting oxygen concentration is the minimum oxygen concentration in a mixture of fuel, air and an inert gas that will propagate flame (Jones [107]). This definition of LOC requires that the flame be observed at some distance away from the ignition source and have traveled through a significant fraction of the enclosed volume (Zlochower [11]). Here λ_1 (or MOC, minimum oxygen concentration) is used to differentiate the flammability from the explosibility, while λ_2 (LOC) is used for representing explosibility or ignitability.

In the diluted flammability diagram, the definition of MOC [108] is

$$\lambda_1 = \lambda(1 - x_{LU}) \tag{5.10}$$

where λ is the ambient oxygen concentration ($\lambda = 0.2095$ for air).

MOC does not have a significant criterion like $R_{LU} = 0$ for LOC, so we have to use Eqs. 5.6 and 5.7 together to get Eq. 5.11 for MOC.

$$\lambda_1 = \lambda(1 - x_{LU}) = \lambda\left(1 - \frac{(1 + R_{LU}) \cdot Q_m}{(1 + R_{LU}) \cdot Q_m - R_{LU} Q_D + C_O H_O - Q_F}\right) \tag{5.11}$$

For normal ambient air, $\lambda = 0.2095$, $Q_m = 1$, we can further simplify this equation as

$$\lambda_1 = \lambda\left(1 - \frac{1 + R_{LU}}{1 + R_{LU} + C_O/\lambda}\right) = \frac{C_O}{1 + R_{LU} + C_O/\lambda} = C_O x_L \tag{5.12}$$

This is the classical correlation for estimating limiting oxygen concentration [97]. Comparing with Eq. 5.9, the difference is small. However, they belong to different families of concepts.

Using above definitions (Eq. 5.9 for LOC, Eq. 5.12 for MOC), we have their responses toward variable oxygen levels (Fig. 5.1, with $Q_D = 0.992$ for N_2 and $Q_D = 1.75$ for CO_2) and different diluents in air (Fig. 5.2, with $\lambda = 0.2095$).

Figure 5.1 shows the response of LOCs to various oxygen levels in air. Though x_{LU} varies sharply in Fig. 5.1a, $\lambda_1 = \lambda(1 - x_{LU})$ has no dependence on oxygen concentration (λ). Equation 5.8 also shows λ_2 has no dependence on oxygen level. It means both are independent of oxygen levels (λ), or they are mainly functions of fuels and diluents in the mixture.

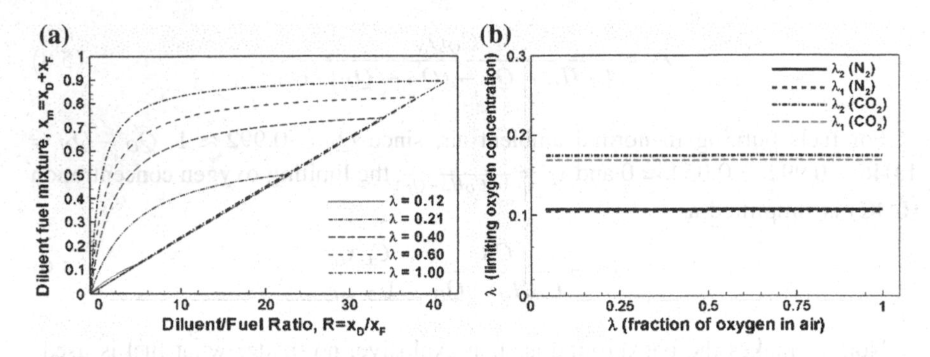

Fig. 5.1 Response of LOCs to variable oxygen levels for propane in air. **a** diluted flammability diagram **b** MOC/LOC responses

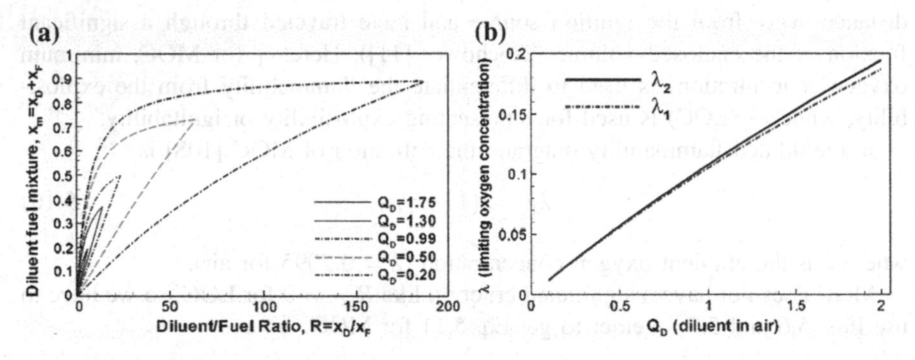

Fig. 5.2 Response of LOCs to variable diluents for propane in air. **a** diluted flammability diagram **b** MOC/LOC responses

However, if we change the diluent type, increasing the quenching potential, LOC and MOC will diverge and show the difference.

Figure 5.2 shows the responses toward various diluents in air. LOC is always larger than MOC for a certain background air, consistent with the estimation expressions in Eqs. 5.9 and 5.12, which are simplified for the case of burning in normal air. The quenching potential of a diluent will change MOC and LOC almost linearly, so the diluent is working more by mass. Since their difference is so small, this may be the reason that Jones [9] claimed that flammability and explosibility are interchangeable. For any fuels burning in air ($\lambda = 0.2095$ and $Q_D = 0.992$), the difference between LOC and MOC is small, only 2.14 % for propane in Fig. 5.2b.

Using the data from a coal mine fire, λ_2 is more fundamental and useful than λ_1 in diagnosing the burning status in such a fire [18]. If the oxygen level drops below λ_2 in the compartment, the flammable envelope disappears or goes to nil. The reason is that λ_2 is a fuel property (explosibility or ignitability), while λ_1 is a process property (flammability), much more complex than it looks ($\lambda_1 = C_O x_L$ for a pure fuel burning in air).

5.1.2 Type II Problem (Variable Diluents in a Stream)

The above discussion solves the case of flammability in a confined space, which is most common for potential hazards of supporting flame propagation. For a type II problem, the risk comes from the ignition potential of a dynamic fuel stream. The purpose of a diluent is mainly to drop the fuel concentration, so the fuel stream is too lean to ignite.

Similar to the previous example, we can set up the conservation of energy at the critical points (LFL and UFL) expressed as

$$\frac{x_L}{1+R}\cdot Q_m+\frac{R\cdot x_L}{1+R}\cdot Q_D+(1-x_L)\cdot 1=\frac{x_L}{1+R}\cdot C_O H_O \tag{5.13}$$

$$\frac{x_U}{1+R}Q_m+\frac{R\cdot x_U}{1+R}Q_D+(1-x_U)\cdot 1=\lambda\cdot(1-x_U)\cdot H_O \tag{5.14}$$

Here $Q_m = Q_F\cdot\beta + Q_D\,(1-\beta)$ is the quenching potential of the diluted fuel. $H_m = H_O$ is unchanged by this dilution, since the fuel type is fixed. The total energy release is scaled down in $C_m = C_O\cdot\beta$.

Solving the above equations, we have the flammable envelope bounded by

$$x_L=\frac{1}{1+\frac{C_mH_m}{1+R}-\frac{Q_m}{1+R}-\frac{Q_DR}{1+R}} \tag{5.15}$$

$$x_U=\frac{\lambda\cdot H_m-1}{\lambda\cdot H_m-1+\frac{Q_m}{1+R}+\frac{Q_D\cdot R}{1+R}} \tag{5.16}$$

Forcing $x_L = x_U$, we have a crossing point (R_{LU}, x_{LU}) which is the theoretical inertion point.

$$R_{LU}=\frac{C_mH_m\lambda-\lambda Q_m-C_m}{\lambda Q_D} \tag{5.17}$$

$$x_{LU}=\frac{1}{1+\frac{C_mH_m}{1+R_{LU}}-\frac{Q_m}{1+R_{LU}}-\frac{Q_DR_{LU}}{1+R_{LU}}} \tag{5.18}$$

Forcing $R_{LU}=0$, or $C_mH_m\lambda-\lambda Q_m-C_m=0$, which leads to

$$\lambda_2=\frac{C_O\beta}{C_O\beta H_O-Q_m} \tag{5.19}$$

Again, the critical MOC is difficult to define, so we have to use Eqs. 5.17, 5.18 and 5.10 combined to solve MOC. The responses of MOC and LOC are explained and shown in Fig. 5.3.

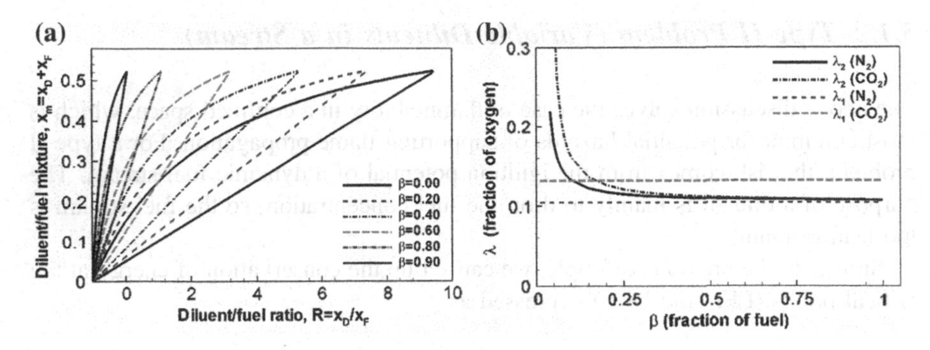

Fig. 5.3 Role of nitrogen on changing the theoretical flammable envelope. **a** diluted flammability diagram **b** MOC/LOC responses

With a variable diluting ratio, the flammable envelopes are modified as shown in Fig. 5.3a. The responses of LOC and MOC are displayed in Fig. 5.3b. It shows that x_{LU} is almost unchanged by the dilution ratio, so MOC is almost constant, independent of the dilution of fuel. MOC is higher using CO_2 for inertion, since CO_2 has a larger molecular weight and a greater quenching potential. The diluent is part of the fuel, turning it into a pseudo fuel burning in background (normal) air. Since the fuel type is same, and the background air is unchanged, MOC stays constant to support flame propagation. Here the contribution of a diluent is realized mainly through dilution by volume.

From Fig. 5.3b, it also shows that the diluting agent will not change LOC until the fuel is almost completely diluted when the diluent begins to influence the background air (by changing its quenching potential). Note LOC is close to MOC of nitrogen-dilution for pure fuels, since nitrogen is the major inerting agent in normal air.

In summary, LOC is more sensitive to fuel fraction, slightly sensitive to diluent type. In contrast, MOC is sensitive to diluent type, but not dependent on fuel quantity. This fact shows that MOC is depending on heat absorption and oxygen depletion, while LOC is dependent on fuel fraction (or the level of dilution).

For a type II problem, there is a critical concept called Limiting Fuel Concentration (LFC) for making the fuel stream non-ignitable. Theoretically, it means the background air cannot supply sufficient oxygen for the diluted fuel to ignite. That means $\lambda_2 = 0.2095$.

Starting with the definition $\lambda_2 = \frac{C_O\beta}{C_O\beta H_O - Q_m} = 0.2095$, we have

$$LFC = \beta_2 = 1 - LDC = \frac{0.2095 \cdot Q_D}{0.2095 \cdot C_O H_O - C_O - 0.2095 \cdot (Q_F - Q_D)} \tag{5.20}$$

where β_2 is the limiting fuel concentration for $\lambda_2 = 0.2095$.

Or assuming $R_{LU} = 0$ for Eq. 5.17, we have the critical or limiting fuel concentration defined as

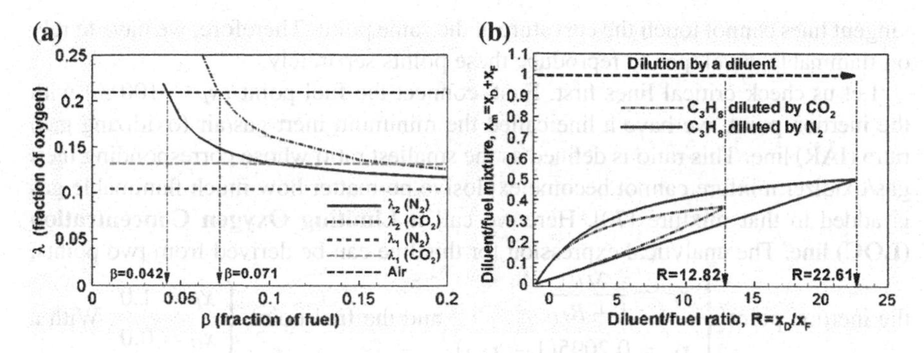

Fig. 5.4 Limiting fuel concentration for diluting a propane stream. **a** LOC curves. **b** Diluted flammability diagram

$$LFC = \beta_2 = \frac{\lambda Q_D}{\lambda C_O H_O - C_O - \lambda \cdot (Q_F - Q_D)} = \frac{1}{1 + R_{LU}} \tag{5.21}$$

Note the original definition of β is the fuel fraction in a mixture, or $\beta = \frac{x_F}{x_F + x_D} = \frac{1}{1+R}$. Therefore, this critical diluting ratio is consistent with the original definition. Submitting parameters for propane/CO_2 mixture, $Q_D = 1.75$, $\lambda = 0.2095$, $C_O = 5$, $H_O = 12.35$, $Q_F = 15.12$, we have the critical fuel concentration as $\beta = 0.072$. That is the limiting fuel concentration with CO_2 dilution. If the fuel fraction drops to this value, ignition needs an oxygen level that the background air cannot supply. It also means the fuel is fully diluted without being ignitable. Similarly, if the diluent is nitrogen ($Q_D = 0.992$), we have the limiting fuel concentration as $\beta = 0.042$. They are demonstrated in Fig. 5.4a, which is a zoom-up view of Fig. 5.3b. The dilution process can also be displayed in Fig. 5.4b, which shows that $\beta_{LU} = \frac{1}{1+R_{LU}} = \frac{1}{1+22.61} = 0.042$ for N_2 dilution, and $\beta_{LU} = \frac{1}{1+R_{LU}} = \frac{1}{1+12.82} = 0.072$ for CO_2 dilution.

This limiting fuel concentration (LFC) is the threshold of fuel being fully diluted even in normal air. This dilution criterion is based on $\lambda_2 > 0.2095$, when the oxygen level in air is insufficient to support ignition. However, it can still be burnt off with a supply of high-temperature air or oxygen-enriched air. This topic will be covered again in Chap. 7.

5.2 Critical Points in a Flammability Diagram

In a theoretical flammability diagram, the inertion point is most important. However, since the experimental curve may not have a clear-cut inertion point, but a curvature, we may have different MIC, MFC, MOC points. Theoretically, these points must be derived from the inertion point. Experimentally, they are different points because the tangent lines cannot touch the curvature at the same point. Therefore, we have to rely on flammable envelopes to reproduce these points separately.

Let us check critical lines first. First, connect the fuel point ($x_F = 100\,\%$) with the inertion point, we have a line called the minimum inert gas/air (oxidizing gas) ratio (IAR) line. This ratio is defined as the smallest ratio whose corresponding inert gas/oxidizer mixture cannot become explosive no matter how much flammable gas is added to that mixture [70]. Here we call it **Limiting Oxygen Concentration (LOC) line**. The analytical expression for this line can be derived from two points, the inertion point $\begin{cases} x_F = \dfrac{x_{LU}}{1+R_{LU}} \\ x_O = 0.2095(1-x_{LU}) \end{cases}$ and the fuel point $\begin{cases} x_F = 1.0 \\ x_O = 0.0 \end{cases}$. With a linear interpolation, we have the LOC line defined as

$$x_O = \frac{0.2095(1-x_{LU})}{1-\frac{x_{LU}}{1+R_{LU}}} \cdot (1-x_F) \tag{5.22}$$

Now we know that $\lambda_1 = 0.2095 \cdot (1-x_{LU})$ and $x_L = \frac{x_{LU}}{1+R_{LU}}$ apply for nitrogen as a diluent ($Q_D = 0.992 \approx 1$), so we have

$$x_O = \frac{\lambda_1}{1-x_L} \cdot (1-x_F) \tag{5.23}$$

Note, $\lambda_1 = C_O \cdot x_L$ and $\lambda_2 = \frac{C_O \cdot x_L}{1-x_L}$, so the LOC line is controlled by

$$x_O = \lambda_2 \cdot (1-x_F) \tag{5.24}$$

That means the critical end point on LOC line is LOC, which is also called In-service Oxygen Concentration (ISOC) [86]. This LOC point means the critical background air is fully inerted and does not support any explosive burning. Therefore, the LOC line is the critical inertion line for background air, which differentiates the explosive/non-explosive states. Under the LOC line, the background air is fully inerted, addition of any more fuel will not support explosive burning.

Similarly, we can draw a line through the air point and tangent to the flammable envelope. The tangent point is another inertion point, Minimal Fuel Concentration (MFC). The cross point with fuel axis is called Limiting Fuel Concentration (LFC). The mathematical expression for this line is

$$x_O = 0.2095 \cdot (1-(1+R_{LU})x_F) \tag{5.25}$$

where the end point of this line on the fuel axis is

$$LFC = \beta_2 = \frac{1}{1+R_{LU}} \tag{5.26}$$

This critical fuel concentration is also called Out-of-service Fuel Concentration (OSFC) [109], or Critical Flammability Ratio (CFR) [110]. Here we call this line as Limiting Fuel Concentration (LFC) line, which is the critical inertion line for defining the non-ignitable state of a fuel stream. Under this LFC line, the fuel stream is fully diluted, so adding more oxygen will not turn the fuel stream ignitable.

The cross point of these two lines is called the inertion point, or the nose point. The iso-oxygen line pass this point is called Minimum Oxygen Concentration (MOC) line. *Here since there are so many important points in the field, those points related to flammability were named after M (maximum in a local view), while those points related to ignitability or explosibility were named after L (limiting in a global view). L is more fundamental than M, in this way, we can quickly recognize their meaning.* This naming rule makes some definitions different from existing ones. However, since existing definitions are already arbitrary and conflicting to each other, we can make them look better by following a rule. These critical points are shown in Fig. 5.5, whereas their definitions are listed in Table 5.1. The analytical expressions for critical lines are provided in Table 5.2.

From Fig. 5.5, we can see that LOC line shows explosibility is fundamental to air, while LFC line shows ignitability is fundamental to fuel. If the fuel

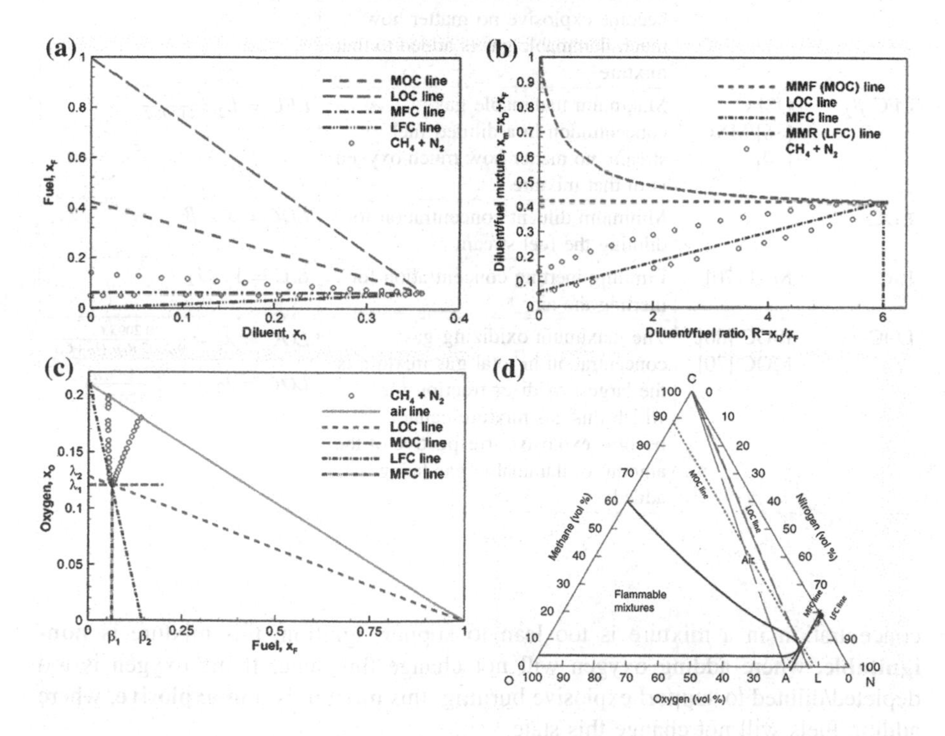

Fig. 5.5 Critical points in flammability diagrams of Methane. **a** Standard flammability diagram. **b** Reduced flammability diagram. **c** Explosive Triangle diagram. **d** Ternary flammability diagram

Table 5.1 The critical concepts in any flammability diagram

Critical points	Old name	Current definition	Relationship to core parameters (R_{LU}, x_{LU})
MFC β_1		Minimal fuel concentration at inertion point	$MFC = \beta_1 = \frac{x_{LU}}{1+R_{LU}}$
MIC	MIC/MAI [70]	Minimum inerting concentration for a specific fuel or for a space	$MIC = x_D = \frac{R_{LU} \cdot x_{LU}}{1+R_{LU}}$
MOC λ_1	LOC	Minimum oxygen concentration for inerting a space for a specific fuel	$MOC = \lambda_1 = 0.2095 \cdot (1 - x_{LU})$
MMR R_{LU}	ICR [70]	Minimal molar (inert/flammable gas) ratio is the smallest ratio whose corresponding inert gas/flammable gas mixture cannot become explosive no matter how much oxidizer is added to that mixture	$MMR = R_{LU}$ $MMR = (1 - \beta_1)/\beta_1$
MMF x_{LU}		Maximum Mixture (diluent+fuel) Fraction is the maximum fraction of the diluent/fuel to be flammable	$MMF = x_{LU}$
MIR	IAR [70]	Minimal inerting (inert gas/air or oxidizing gas) ratio (MIR) is the smallest ratio whose corresponding inert gas/oxidizer mixture cannot become explosive no matter how much flammable gas is added to that mixture	$MIR = \frac{x_D}{x_O} = \frac{\frac{R_{LU}}{1+R_{LU}} \cdot x_{LU}}{0.2095 \cdot (1-x_{LU})}$ $MIR = (1 - \gamma_1)/\gamma_1$
LFC β_2	OSFC [86]/MXC [70]	Maximum flammable gas concentration in a diluted fuel stream, no matter how much oxygen is in that mixture	$LFC = \beta_2 = \frac{1}{1+R_{LU}}$
LDC		Minimum diluent concentration for diluting the fuel stream	$LDC = 1 - \beta_2$
LIC	MAI [70]	Limiting inerting concentration for inerting the air	$LIC = 1 - \lambda_2$
LOC λ_2	ISOC [86]/ MOC [70]	The maximum oxidizing gas concentration in total gas mixture is the largest oxidizer fraction for which this gas mixture cannot become explosive irrespective of the amount of flammable gas being added	$LOC = \lambda_2 = \frac{0.2095 \cdot C_O}{0.2095 \cdot R_{LU} \cdot Q_D + C_O}$ $LOC = \lambda_2 = \frac{0.2095(1-x_{LU})}{1-\frac{x_{LU}}{1+R_{LU}}}$

concentration in a mixture is too lean to support ignition, this mixture is **non-ignitable,** where adding oxygen will not change this state. If the oxygen is too depleted/diluted to support explosive burning, this mixture is **non-explosive**, where adding fuels will not change this state.

Table 5.2 The analytical threshold lines in any flammability diagram

	Standard	Diluted	Triangle
Air line, AB	$x_D = 0$	$R = 0$	$x_O = 0.2095 \cdot (1 - x_F)$
LFC line	$x_D = R_{LU} \cdot x_F$	$R = x_D / x_F = R_{LU}$	$x_O = 0.2095 \cdot [1 - (1 + R_{LU})x_F]$
MOC line	$x_D + x_F = 1 - 4.773 \cdot \lambda_1$	$x_m = 1 - 4.773 \cdot \lambda_1$	$x_O = \lambda_1$
LOC line	$x_F = \frac{x_{LU} - 1 - R_{LU}}{R_{LU} \cdot x_{LU}} \cdot x_D + 1$	$x_m = \frac{(1 - 4.773 \cdot \lambda_2) \cdot (1 + R)}{1 + R - 4.773 \cdot \lambda_2}$	$x_O = \lambda_2 (1 - x_F)$
MFC line	$x_F = \beta_1 = \frac{x_{LU}}{1 + R_{LU}}$	$x_m = \frac{x_{LU} \cdot (1 + R)}{1 + R_{LU}}$	$x_F = \beta_1 = \frac{x_{LU}}{1 + R_{LU}}$
Stoichiometric line	$x_F = x_{st,0} \cdot (1 - x_D)$	$x_m = \frac{x_{st,0} \cdot (1 + R)}{1 + x_{st,0} \cdot R}$	$x_O = \frac{1 - x_{st,0}}{4.773 \cdot x_{st,0}} \cdot x_F$

MOC line is passing the cross point (the inertion point) of LOC and LFC lines, so flammability is a special case depending on both LOC and LFC lines. Thus, the flammability is not a fundamental property of fuel or oxygen, but a result of both. If the oxygen level drops below MOC line, the mixture will not support flame propagation, or the mixture is **non-flammable**. Non-flammable is metastable state, since adding fuel or oxygen will tip the balance easily.

5.3 Critical Parameters for a Flammability Diagram

In the diluted diagram (Fig. 5.5b), the inertion point is represented as (MMR, MMF). Since MMR and MMF play a fundamental role on determining other critical points or lines, we can determine their values from the flammability diagrams in advance. These two pieces of information comprise a parameter set sufficient for diluting and purge operations. Thus, graphical operations on a flammability diagram can be waivered if we have such a parameter database. Table 4.2 lists these critical parameters taken from the diagrams published in those famous reports by Coward [4] and Zabetakis [43]. Theoretical MMR and MMF are determined by Eqs. 5.6 and 5.7. For most fuels, theoretical MMR and MMF are larger than their counterparts in experimental values (see Table 4.2), giving the theoretical treatment a conservative nature.

Here we have a new look at these two concepts. From the original definition, $\lambda_1 = \lambda \cdot (1 - x_{LU})$, we have

$$x_{LU} = 1 - \frac{\lambda_1}{\lambda} = \frac{\lambda - \lambda_1}{\lambda} \tag{5.27}$$

So MMF (x_{LU}) is **the fraction of excess background oxygen in reference to Minimal Oxygen Concentration (MOC)**. MMF belongs to the concept of flammability.

Rearrange the terms in Eq. 5.16, we have

$$R_{LU} = \frac{C_O H_O \lambda - \lambda Q_F - C_O}{\lambda Q_D} = \frac{C_O H_O - Q_F}{Q_D} - \frac{C_O}{\lambda Q_D} = \frac{C_O}{Q_D}\left(\frac{1}{\lambda_2} - \frac{1}{\lambda}\right) \tag{5.28}$$

So the physical meaning of MMR (R_{LU}) is **the amount of diluent with Q_D to absorb the excess heating potential of air** (C_O/λ) **above needed** (C_O/λ_2). MMR belongs to the concept of explosibility.

From Eqs. 5.27 and 5.28, we can see that x_{LU} controls flammability, while R_{LU} controls both explosibility and ignitability.

From Eq. 5.28, we can have another expression for experimental LOC

$$\lambda_2 = \frac{0.2095 \cdot C_O}{0.2095 \cdot R_{LU} \cdot Q_D + C_O} \tag{5.29}$$

For any mixture safety problem, if MMR and MMF of a fuel are provided like those in Table 4.2, we can directly derived the operation curves and critical points using the equations in Table 5.1. If the fuel is not listed in Table 4.2, we can use the theoretical equations to estimate their values. If the diluent is not nitrogen, we can use the molecular ratio (MW/28) to rescale the equations (Eqs. 4.57 and 4.58), then used rescaled R_{LU} and x_{LU} to find critical parameters for safe operations. This provides a framework of methodology in safe handling of flammable gases.

5.4 Two Modes of Suppression

Generally, there are two types of problems dealing with mixture safety. Type I problem deals with a static mixture in a confined space. Change of one component (such as fuel) will affect other components (such as oxygen and diluents) accordingly. The dynamics process for updating all concentration profiles in a compartment is called purge, with the purpose to inertize the compartment against any flame propagation. A typical type I problem is the compartment fire, where the oxygen concentration is fluctuating over time. The counter-solution to this problem is **the total flooding** of gaseous agents, which is also a type I problem. Without fuel, the addition of diluent will inertize the air (oxygen), which is an effective means against explosions, including dust explosions.

Type II problem deals with a dynamic fuel stream, usually composed of a fuel and a diluent, releasing into air. Change of one component (such as the fuel) will affect the accompanying constituent (such as the diluent), but not the background air (or oxygen). The dynamic process to change the fuel/diluent ratio is called dilution, while the background (or oxygen) condition stays the same. A typical type II problem is the inertion of a fuel leak, where the fuel concentration is so diluted that the ignition is not possible in its diluted state. Similarly, **the local application** of gaseous agents plays the same role on the fuel, so it is also a type II problem.

For a type I problem, we can demonstrate the inertion process via oxygen-modified flammability diagram, which is also derived from the Heating-Quenching Ratio (HQR) diagram.

For a type II problem, the adding of diluent will change the thermal signature of the fuel, thereby modifying the flammability diagram as shown in Fig. 5.7.

Figure 5.6 shows the impact of applying diluent in a total flooding, while Fig. 5.7 shows the response of flammability diagram in a local application of diluent.

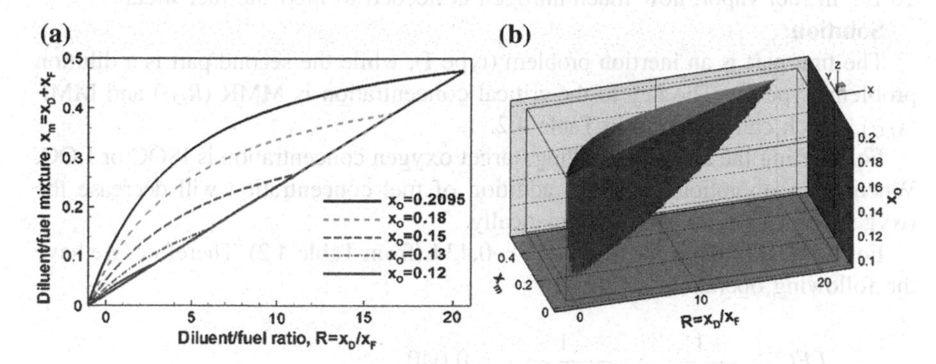

Fig. 5.6 Flammability diagrams for total flooding (R290+N_2) [111]. **a** Diluted flammability diagram. **b** HQR diagram for HQR = 1 envelope

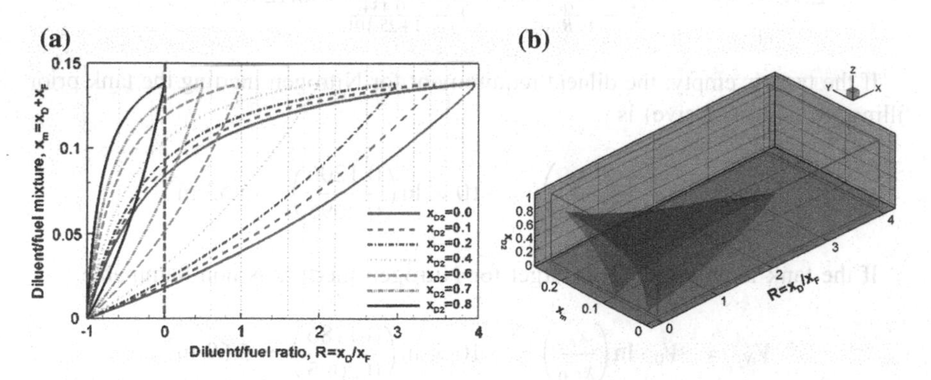

Fig. 5.7 Flammability diagrams for local application (R290+R227ea). **a** Diluted flammability diagram. **b** HQR diagram for HQR = 1 envelope

5.5 Problems and Solutions

5.5.1 Tank Safety

Here is one example to explain the utility of MMR/MMF in dilution/purge operations.

Problem 5.1 Historically, AVgas tank in an aircraft carrier is protected by sea-water. That is, if 100 gallon of Aviation Gasoline is needed, 100 gallons of water has to be pumped into the storage tank, so there is no flammable space with dangerous oxygen inside [112]. Now, how to inert the head space if nitrogen is used to protect a headspace of 10 m^3? If the pipeline to deliver Avgas has a leakage with 10 L/s in fuel vapor, how much nitrogen is needed to inert the fuel stream?

Solution:

The first part is an inertion problem (type I), while the second part is a dilution problem (type II). The key to the critical concentration is MMR (R_{LU}) and MMF (x_{LU}), which can be found in Table 4.2.

For inerting the air prior to filling, target oxygen concentration is ISOC or LOC. When the evaporation starts, the addition of fuel concentration will decrease the oxygen from LOC to MOC automatically.

For AVgas, MMR = 24.101, MMF = 0.434 (from Table 4.2). Therefore, we have the following operation targets:

$$LFC = \frac{1}{1 + R_{LU}} = \frac{1}{1 + 24.101} = 0.040$$

$$MOC = \lambda_1 = 0.2095 \cdot (1 - x_{LU}) = 0.2095 \times (1 - 0.434) = 0.1186$$

$$LOC = \lambda_2 = \frac{0.2095(1 - x_{LU})}{1 - \frac{x_{LU}}{1+R_{LU}}} = \frac{0.1186}{1 - \frac{0.434}{1+25.101}} = 0.1206$$

If the tank is empty, the diluent requirement for Nitrogen inerting the tank prior filling (to non-explosive) is

$$V_{N_2} = -V_0 \cdot \ln\left(\frac{x_F}{x_{F,0}}\right) = -10 \times \ln\left(\frac{0.1206}{0.2095}\right) = 5.52 \text{ m}^3$$

If the tank is not empty, the target for Nitrogen inerting is non-flammable,

$$V_{N_2} = -V_0 \cdot \ln\left(\frac{x_F}{x_{F,0}}\right) = -10 \times \ln\left(\frac{0.1186}{0.2095}\right) = 5.69 \text{ m}^3$$

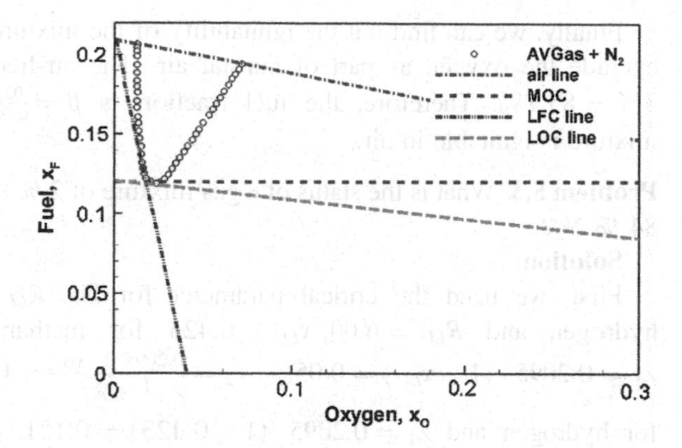

Fig. 5.8 The explosive triangle for Avgas with nitrogen dilution

The diluent requirement for Nitrogen diluting the fuel stream (to non-ignitable) is

$$\dot{V}_{N_2} = \frac{\dot{V}_{fuel}}{LFC} - \dot{V}_{fuel} = \frac{10}{0.040} - 10 = 240 \text{ L/s}$$

These critical points can be identified from the explosive triangle diagram below (Fig. 5.8).

5.5.2 Mixture Flammability

Here are three problems with mixture flammability. Note explosibility and ignitability are stressed for two types of problems.

Problem 5.2 What is the status of a gas mixture of 5 % H_2, 3 % O_2, and 92 % N_2?

Solution:

First, we need the critical parameter for H_2, $R_{LU} = 16.42$, $x_{LU} = 0.76$ for hydrogen. Then we have

$$\lambda_1 = 0.2095 \cdot (1 - x_{LU}) = 0.05$$

$$\lambda_2 = \frac{0.2095 \cdot (1 - x_{LU})}{1 - \frac{x_{LU}}{1+R_{LU}}} = 0.052$$

$$\beta_2 = \frac{1}{1 + R_{LU}} = 0.057$$

Secondly, we can find out the explosibility of the fuel-free mixture. Here the fuel-free mixture is 3 % O_2 plus 92 % N_2, or $\lambda = \frac{0.03}{0.03+0.92} = 0.032 < \lambda_2$. Therefore, the background air is non-explosive for flame propagation.

Thirdly, we can find the flammability of the mixture, which is $\lambda = 0.03 < \lambda_1$. Therefore, the mixture is non-flammable.

Finally, we can find out the ignitability of the mixture. Before that, we need to exclude the oxygen as part of normal air. The air-free mixture is $1 - 4.773 \times 3\,\% = 85.7\,\%$. Therefore, the fuel fraction is $\beta = \frac{0.05}{0.857} = 0.058 > \beta_2$. So this mixture is ignitable in air.

Problem 5.3 What is the status of a gas mixture of 2 % H_2, 1 %CH_4, 13 % O_2, and 84 % N_2?

Solution:

First, we need the critical parameter for H_2, $R_{LU} = 16.42$, $x_{LU} = 0.76$ for hydrogen and $R_{LU} = 6.00$, $x_{LU} = 0.423$ for methane. Then we will have $\lambda_1 = 0.2095 \cdot (1 - x_{LU}) = 0.05$, $\lambda_2 = \frac{0.2095 \cdot (1 - x_{LU})}{1 - \frac{x_{LU}}{1 + R_{LU}}} = 0.052$, $\beta_2 = \frac{1}{1 + R_{LU}} = 0.057$ for hydrogen and $\lambda_1 = 0.2095 \cdot (1 - 0.423) = 0.121$, $\lambda_2 = \frac{0.2095 \cdot (1 - x_{LU})}{1 - \frac{x_{LU}}{1 + R_{LU}}} = 0.129$, $\beta_2 = \frac{1}{1 + R_{LU}} = 0.143$ for hydrogen.

Secondly, we can find out the explosibility of the fuel-free mixture. Here the fuel-free mixture is 13 % O_2 plus 84 % N_2, or $\lambda = \frac{0.13}{0.13 + 0.84} = 0.134 > \lambda_2$ for either fuel. Therefore, the background air is explosive for flame propagation.

Thirdly, we can find the flammability of the mixture, which is $\lambda = 0.134 > \lambda_1$ for either fuel. So the mixture is self-flammable.

Finally, we can find out the ignitability of the mixture. Before that, we need to exclude the oxygen as part of normal air. The air-free mixture is $1 - 4.773 \times 13\,\% = 35.0\,\%$, which includes nitrogen and fuel. Now 2 % hydrogen needs $x_D = \frac{x_F}{\beta_2} = \frac{0.02}{0.057} = 35.1\,\%$ to achieve non-ignitable, while 1 % methane needs nitrogen $x_D = \frac{x_F}{\beta_2} = \frac{0.01}{0.143} = 7.0\,\%$ to achieve non-ignitable. Altogether, 2 % H_2 and 1 % CH_4 need 35.1 + 7 % to get full-dilution, while the available nitrogen for fuel dilution is 35.0 %−2 %−1 % = 32 %. Therefore, there is insufficient nitrogen in the mixture to achieve non-ignitable. This mixture is ignitable in air.

Problem 5.4 What is the status of a gas mixture of 10 % CO, 5 % O_2, 20 % CO_2, 25 % Ar and 30 % Ne?

Solution:

This is a problem of multiple diluents. First, we need to convert other diluents into nitrogen-equivalent. Here the quenching potential for N_2, CO_2, Ar, Ne are 0.992, 1.75, 0.65, 0.65 respectively. The equivalent nitrogen for this mixture is $20\,\% \cdot \frac{1.75}{0.992} + 25\,\% \cdot \frac{0.65}{0.992} + 25\,\% \cdot \frac{0.65}{0.992} = 68\,\%$. So the new equivalent mixture is 10 % CO, 5 % O_2 and 68 % N_2.

Now, retrieve the critical parameters for CO from Table 4.2, $R_{LU} = 4.08$, $x_{LU} = 0.73$ for hydrogen. Then we have

$$\lambda_1 = 0.2095 \cdot (1 - x_{LU}) = 0.0566$$

$$\lambda_2 = \frac{0.2095 \cdot (1 - x_{LU})}{1 - \frac{x_{LU}}{1+R_{LU}}} = 0.0661$$

$$\beta_2 = \frac{1}{1 + R_{LU}} = 0.197$$

For the fuel-free mixture, the oxygen concentration is 5 % O_2 plus 68 % N_2, or $\lambda = \frac{0.05}{0.05+0.68} = 0.0685 > \lambda_2$. Therefore, the background air is explosive for flame propagation.

For the flammability of the mixture, $\lambda = \frac{0.05}{0.05+0.10+0.68} = 0.060 > \lambda_1$, so the mixture is flammable.

Finally, we can find out the ignitability of the mixture. Before that, we need to exclude the oxygen as part of normal air. The air-free fraction of diluent is $68\,\% - 3.773 \times 5\,\% = 49.1\,\%$. Therefore, the fuel fraction is $\beta = \frac{0.1}{0.491+0.1} = 0.169 < \beta_2$. So this mixture is non-ignitable in air.

Problem 5.5 Let us redo the problem in Sect. 2.3. A gaseous mixture of 7 % of H_2 in CO_2, is it flammable in air? [71]

Solution:

This is a typical type II problem. From Table 4.2, we have the critical parameter for Hydrogen inerted by Nitrogen is $MMR = R_{LU} = 16.42$. Then the critical parameter for Hydrogen inerted by Carbon Dioxide is $MMR = \frac{R_{LU}}{Q_D} = \frac{16.42}{1.75} = 9.38$. The Limiting Fuel Concentration is

$$LFC = \frac{1}{1 + R_{LU}} = \frac{1}{1 + 9.38} = 0.096$$

Here $x_F = 0.07 < LFC = 0.096$, so the mixture is still in its fully diluted state and non-ignitable.

Another solution:

Find the thermal signature of Hydrogen first

$$H_O = \frac{x_U - x_L}{C_O \cdot x_U \cdot x_L - x_O \cdot (1 - x_U) \cdot x_L} = \frac{0.75 - 0.04}{0.5 \times 0.75 \times 0.04 - 0.2095 \times (1 - 0.75) \times 0.04} = 55.01$$

$$Q_F = 1 - 1/x_L + C_O H_O = 1 - 1/0.04 + 0.5 \times 55.01 = 3.521$$

Then apply Eq. 5.19, we have

$$\lambda_2 = \frac{\beta \cdot C_O}{\beta \cdot H_O \cdot C_O - Q_m} = \frac{0.07 \times 0.5}{0.07 \times 55.01 \times 0.5 - (0.93 \times 1.75 + 0.07 \times 3.521)}$$
$$= 0.681 > 0.2095$$

Since the required oxygen level (ignitability) is more than the background air could supply, this mixture is non-ignitable.

Problem 5.6 What is the limiting oxygen concentration (LOC, or In-service oxygen concentration, ISOC) for methane and hydrogen?

Solution:

$$\lambda_2 = \frac{0.2095 \cdot C_O}{0.2095 \cdot R_{LU} \cdot Q_D + C_O} = \frac{0.2095 \times 2}{0.2095 \times 5.77 \times 0.992 + 2} = 0.131$$

$$\lambda_2 = \frac{0.2095 \cdot C_O}{0.2095 \cdot R_{LU} \cdot Q_D + C_O} = \frac{0.2095 \times 0.5}{0.2095 \times 16.42 \times 0.992 + 0.5} = 0.0472$$

Experimentally, Ishizuka and Tsuji [113] got 0.143 and 0.052, while Simmons and Wolfhard got 0.139 and 0.054 [114] respectively. Their designs of difusion flame provide an effective platform for the measurement of explosiblity and ignitability.

Chapter 6
Operations Within Flammability Diagrams

Thermal balance method is universal in determining the flammable state of a sample mixture, but difficult to treat various possibilities arising from dynamic changes in composition. In addition, there are many cases dealing with safe operations, how to turn a flammable mixture safe? In terms of trend demonstration, a diagram is useful to present the inter-relationship between each variable.

With clearly defined concepts of ignitability, flammability and explosibility, it is possible to draw the operation routes to reach safe states in a flammability diagram. For each diagram, there is a critical oxygen line (LOC line), and a critical fuel line (LFC) line. Flammability occurs near the cross point of both lines.

For a type I problem, the adding diluent is to control the explosibility (oxygen) first, then the ignitability (fuel). Since all three species concentrations have to be changed simultaneously, a purge operation is needed to reach this goal.

For a type II problem, the purpose is to control the ignitability of a fuel stream, while the background explosibility (oxygen) is always fixed. Since any fuel concentration in a fuel stream can be controlled as a result of diluent concentration change, a dilution operation is typical for type II problems.

Through analytical expressions for purge and dilution, we can define the route to safety without the help of a diagram, just solving the governing equations simultaneously to find critical points discussed in Sect. 5.2.

6.1 Flammable States in Diagrams

A flammability diagram can be divided into different zones by some critical lines. The position of the sample point shows the flammable state of a mixture. Let us start with the diluted flammability diagram (Fig. 6.1), where MMR and MMF lines encompass a rectangular domain. Firstly, a stoichiometric reaction line divides the flammable domain into a fuel-rich and a fuel lean zone. Then the LFL and UFL lines isolate a small area called flammable zone. Above this zone, the mixture is too rich to react. Below this zone, the mixture is too lean to react. Then there is another two lines, MMF (MOC) line divides unconditional non-flammable (above MMF line)

T. Ma, *Ignitability and Explosibility of Gases and Vapors*,
DOI 10.1007/978-1-4939-2665-7_6

zone and conditional flammable (below MMF line, but above UFL line) zone. Conditional means the mixture will still be flammable if the fuel/diluent ratio is changed without a change in oxygen. Unconditional means the mixture is non-flammable even if the fuel fraction is changed. Similarly, there is a MMR (LFC) line, which isolates further a space called non-ignitable. Non-ignitable means the fuel concentration is too diluted to support ignition, no matter how much oxygen is involved. In-between MMF and MMR lines, there is a LOC line, which defines a zone called non-explosive. Combinations of MMR/MMF/LOC and LFL/UFL lines divide the space into six zones of interest, flammable, rich, lean, non-flammable, non-explosive and non-ignitable. These six zones are demonstrated in Fig. 6.1.

Operation points in these zones may have different levels of safety. If the operation point moves just outside of the flammable domain, either in rich or lean domain, it has **a minimal safety margin**. If the diluent level or fuel fraction is changed, it may moves back into flammable zone again with added fuel or oxidizer.

If the operation point moves above the MMF (MOC) line into the non-flammable domain, it has **an intermediate safety margin**. Theoretically, non-flammable means $\lambda < \lambda_1$. The mixture is not supporting the flame propagation due to lacking of flammability (oxygen). Adding diluent/fuel will not move the operation point below the MOC line, unless air (oxygen) is added into the mixture, bringing in the needed flammability.

The concept of explosibility is similar to flammability, but controlled by the LOC line. Theoretically, non-explosiveness means $\lambda < \lambda_2$. For most of the time, explosibility is harder to control than flammability ($\lambda_1 > \lambda_2$), with a small exception near the inertion point, where $\lambda_1 < \lambda_2$, meaning non-explosibility is easier to achieve when the fuel concentration is too small. The MMF point is the cross point of MMF and LOC lines, meaning $\lambda_1 = \lambda_2$. Historically, this concept is not fully recognized, so no operation is performed based on the explosibility (or LOC) line. They are assumed interchangeable, since $\lambda_1 = \lambda_2$ at the MMF point.

If the operation point moves further to the right of the MMR (LFC) line into the non-ignitable zone, it has **a highest safety margin**. The fuel is too lean to ignite, and adding air will not turn the mixture ignitable. The fuel or mixture is said to be

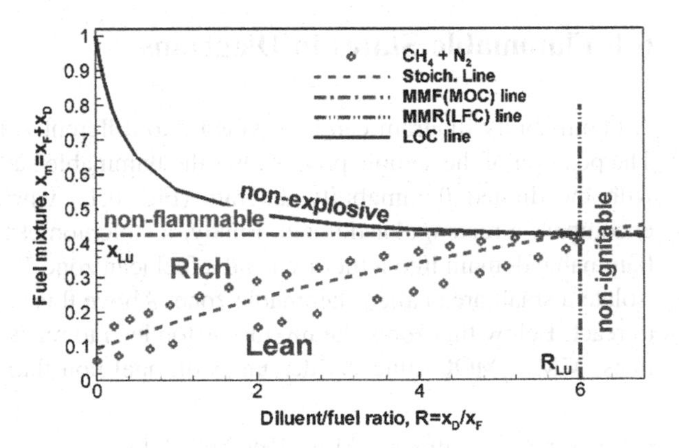

Fig. 6.1 Flammable zones and lines in a diluted flammability diagram for methane

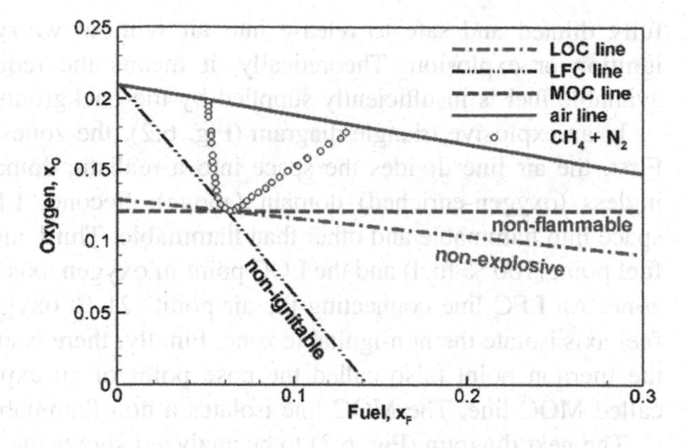

Fig. 6.2 Flammable zones and lines in an explosive triangle diagram for methane

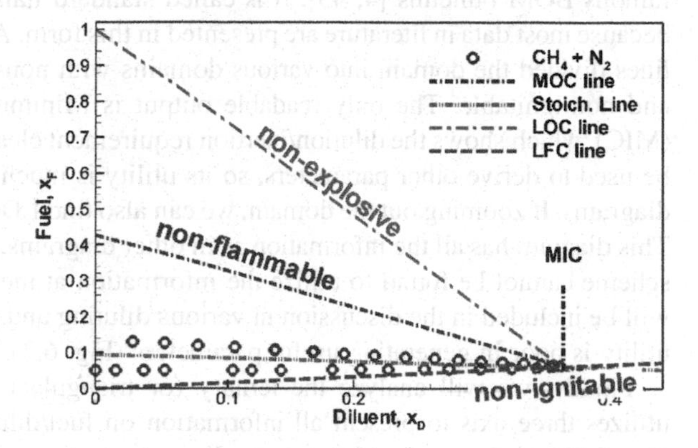

Fig. 6.3 Flammable zones and lines in a standard flammabilty diagram for methane

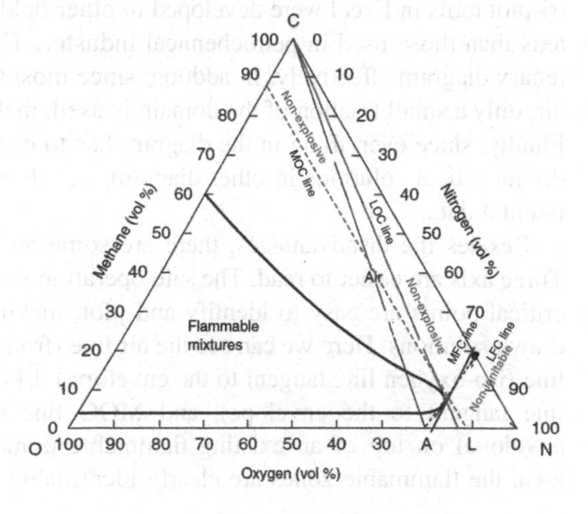

Fig. 6.4 Flammable zones and lines in a ternary flammability diagram for methane

fully diluted and safe to release into air without worrying about the danger of ignition or explosion. Theoretically, it means the required LOC to match the available fuel is insufficiently supplied by the background air, or $\lambda_2 > 0.2095$.

In an explosive triangle diagram (Fig. 6.2), the zones are not in regular shape. First, the air line divides the space into a realistic domain (below), and a meaningless (oxygen-enriched) domain (above). Second, LFL/UFL lines divide the space into flammable and other than flammable. Third, an LOC line connecting the fuel point (100 % fuel) and the LOC point in oxygen axis isolates the non-explosive zone. An LFC line connecting the air point (21 % oxygen) and the LFC point in fuel axis isolate the non-ignitable zone. Finally, there is an iso-Oxygen line passing the inertion point (also called the nose point of an explosive triangle), which is called MOC line. The MOC line isolates a non-flammable zone.

The next diagram (Fig. 6.3) to be analyzed shows the role of diluent on inerting fuels, which is most convenient for presenting experimental data, especially in the famous BOM bulletins [4, 43]. It is called standard flammability diagram, simply because most data in literature are presented in this form. Again, the MOC/LOC/LFC lines divided the domain into various domains with non-flammable, non-explosive and non-ignitable. The only readable output is Minimum Inerting Concentration (MIC), which shows the dilution/inertion requirement clearly. However, MIC cannot be used to derive other parameters, so its utility is much lower than the above two diagrams. If zooming out the domain, we can also find LDC/LFC point, or LOC/LIC. This diagram has all the information as in other diagrams, while a simple conversion scheme cannot be found to utilize the information at inertion. For completeness, it will be included in the discussion in various diluting and purge operations, while its utility is poor in generating useful parameters (Fig. 6.3).

Finally, we will analyze the ternary (or triangular) diagram (Fig. 6.4) which utilizes three axis to present all information on fuel/diluent/oxidzer compositions. The ability to draw a 5-points ternary diagram is recommended for any textbook on process safety, while a standard format to present data can hardly be found. Most tri-plot tools in Excel were developed in other fields, which has a different layout of axis than those used in petrochemical industry. This is the major obstacle to use a tenary diagram effectively. In addtion, since most flammability problems happen in air, only a small fraction of the domain is used, making estimations rough in values. Finally, since everything in the diagram has to convert coordinates before plotting, the analytical solutions in other diagrams are difficult to plot on top of the experimental data.

Besides the disadvantages, there are some advantages to use ternary diagram. Three axis are easier to read. The safe operation routes are simple to draw, while the critical points are easy to identify and plot, making it an ideal platform for hand-drawn solutions. Here we can see the air line (from the fuel point to air point), LOC line (iso-oxygen line tangent to the envelope), LFC line (from the air point, draw a line tangent to the envelope), and MOC line (iso-oxygen line tangent to the envelope) on top of an existing flammable domain. Within the limited effective area, the flammable zones are clearly identifiable with the above threshold lines.

From the above analysis, the flammable zones are equivalent to each other in all four diagrams. Except the ternary diagram, they are isolated by characteristic lines represented by analytical solutions listed in Table 5.2. Even we have three levels of flammability (non-flammable, non-explosive, non-ignitable), the state of non-explosiveness was not properly recognized and defined in the past, so the target of dilution operation is either non-flammable, or non-ignitable. Here we will work on graphical representation of operate routes to reach such safe targets.

6.2 A Thermal Explanation

Here we can have a thermal check on flammability diagrams. Similar to the energy ratio proposed in Eqs. 4.12 and 4.13, we have two Heating-Quenching Ratios (HQR) defined as

$$HQR_1 = \frac{C_O H_O x_F}{Q_F x_F + Q_O x_O + Q_D x_D} = \frac{\text{fuel heating potential}}{\text{mixture quenching potential}} \tag{6.1}$$

$$HQR_2 = \frac{H_O \cdot x_O}{Q_F x_F + Q_O x_O + Q_D x_D} = \frac{\text{oxidizer heating potential}}{\text{mixture quenching potential}} \tag{6.2}$$

If $HQR_1 > 1$, the fuel heating potential is more than the mixture quenching potential, which means fuel is sufficient to burn, or $x > x_L$. It also means the fuel in the mixture is ignitable, or the mixture is ignitable.

If $HQR_2 < 1$, the oxygen heating potential is more than the mixture quenching potential, which means oxygen is sufficient to burn, or $x < x_U$. It also means the air in the mixture is explosive, or the mixture is explosive.

The total energy release is limited by both fuel and oxygen, or the smaller of the two, as shown in Eq. 6.3.

$$HQR = \min(HQR_1, HQR_2) \tag{6.3}$$

The distribution of HQR in the diluted flammability diagram is shown in Fig. 6.5. If $HQR > 1$, the fuel or oxygen-based heating potential is more than the mixture quenching potential, which means the mixture is both ignitable and explosive, more precisely flammable, or $x_L < x < x_U$. So the flammable envelope is controlled by $HQR = 1$.

Figure 6.5 shows the distribution of various HQRs for methane burning in normal air. The flammable envelope covers the region where the heating potential is larger than the quenching potential of the mixture. Using HQR = 1 to get the iso-contour line, the resulting peninsula is; in the diluted flammability diagram.

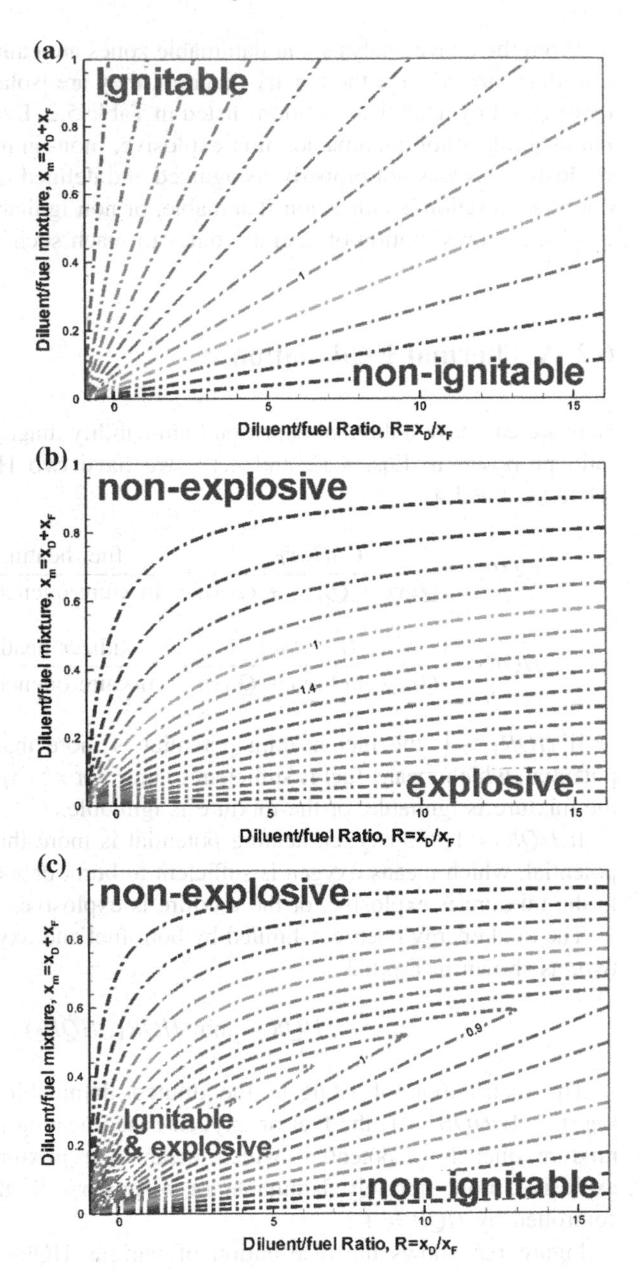

Fig. 6.5 The distribution of HQR in the flammability domain for methane burning in air. **a** Contour of HQR_1 (Fuel-dominated energy release), **b** Contour of HQR_2 (Oxygen-dominated energy release), **c** Contour of $HQR = \min(HQR_1, HQR_2)$

By manipulating the cumulative quenching and heating potentials in a mixture, it is possible to find an iso-HQR surface in a 3-D HQR diagram. For the HQR = 1 surface, it is also the surface for critical (upper and lower) flammable limits. Here is a 3-D HQR surface modified by oxygen fraction in the mixture (Fig. 6.6). Governing equations and variations are provided in Sect. 7.2.1.

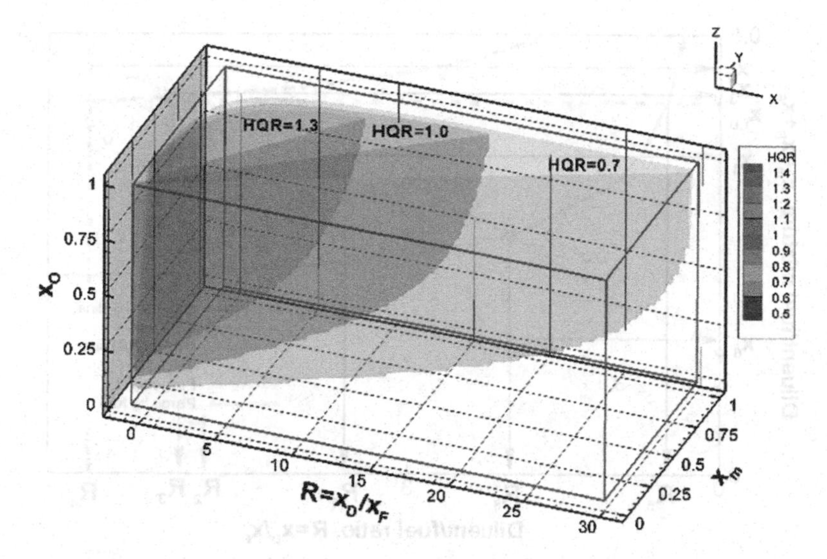

Fig. 6.6 The contour surface of oxygen-modified flammable envelope for methane

Figure 6.6 shows the distribution of HQR for the combustion of a CO_2-diluted fuel (Methane) in oxygen-modified air. HQR = 1 is the contour surface for the flammable envelope.

6.3 Purge in Flammability Diagrams

6.3.1 Different Levels of Safety

We have three levels of safety margin with three points as targets in the flammability diagram, which means three targets for safe operations. A diluent adding to a flammable mixture will move the operation point to targets at three levels, out-of-envelope, non-flammable and non-ignitable (as shown in Fig. 6.7). Here in order to understand such three levels of safety, we need to start with a sample flammable mixture.

Here is a sample taken from a sealed area with the following mixture composition: CH_4–9 %, N_2–57.7 %, O_2–15.3 %. Is this gas-mixture flammable? What are the possible safe targets of operations?

Here the diluent (nitrogen) level for inertion is $1 - 0.09 - 4.773 * 0.153 = 0.18$. So $R = 0.18/0.09 = 2$, $x_m = 0.18 + 0.09 = 0.27$. We can plot the experimental flammability diagram for methane as shown in Eq. 6.7, where the sample lies in the up-middle part in the flammable envelope. Therefore, this sample is flammable. Since such a typical type I problem is difficult to control two species at the same time (by dilution), purge is resorted to manipulate concentrations to reach the safe state. However, we have three choices for purge, purge by air, purge by diluent and purge

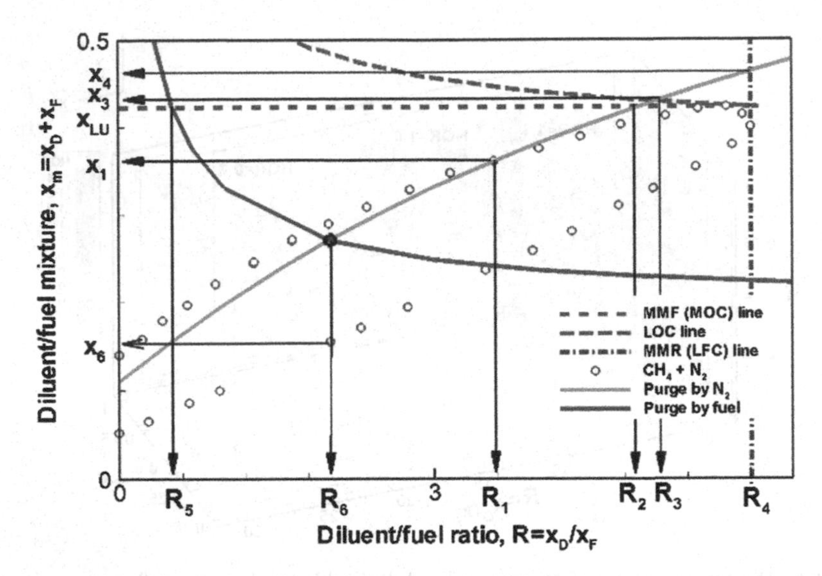

Fig. 6.7 Purge operations on the flammable mixture from a mine fire

by fuel. Their purge curves (with analytical expressions provided in Table 6.1) are displayed in Figs. 6.8, 6.9, 6.10 and 6.11.

If purged with air, the sample point should move down to (R_6, x_6), to reach the lean state.

If purged with fuel (methane), the sample point will move up to (R_5, x_{LU}), to reach the non-flammable state.

If purged with diluent (nitrogen), the sample point will first move to (R_1, x_1), to reach **the fuel-rich state**. However, this is not safe enough. So the purge should stop at least near (R_2, x_{LU}), to reach **the non-flammable state**. Since the non-flammable state still has the danger of crossing the flammable zone when diluted by air, the purge should stop at (R_3, x_3), to reach **the non-explosive state**. Finally, the purge should stop at (R_4, x_4) to reach **the non-ignitable** state. After that, the mixture is both non-explosive (oxygen-lean, no matter how much fuel is available) and non-ignitable (fuel-lean, no matter how much oxygen in available), so it is safe.

Since the safety targets are clearly defined, we can draw the purge routes in these diagrams. For the purge operations with diluent/fuel/air, we have the following dilution curves provided in Table 6.2. All curves are functions of initial condition

Table 6.1 Analytical expressions for purge operations

	Standard	Diluted	Triangle
Purge (D)	$x_F = \frac{1-x_D}{1-x_{D,0}} \cdot x_{F,0}$	$x_m = \frac{x_{F,0}\cdot(1+R)}{R\cdot x_{F,0}+1-x_{D,0}}$	$x_O = 0.2095 \cdot \left(1 - x_{F,0} - x_{D,0}\right) \cdot \frac{x_F}{x_{F,0}}$
Purge (F)	$x_F = 1 - \frac{1-x_{F,0}}{x_{D,0}} \cdot x_D$	$x_m = \frac{x_{D,0}\cdot(1+R)}{R\left(1-x_{F,0}\right)+x_{D,0}}$	$x_O = 0.2095\left(1 - x_{F,0} - x_{D,0}\right) \cdot \frac{1-x_F}{1-x_{F,0}}$
Purge (A)	$x_F = x_D \cdot x_{F,0}/x_{D,0}$	$R = R_0$	$x_O = 0.2095\left[1 - \left(1 + \frac{x_{D,0}}{x_{F,0}}\right) \cdot x_F\right]$

Table 6.2 Analytical expressions for diluting operations

	Standard	Diluted	Triangle
Dilution (D/A)	$x_F = x_{F,0}$	$x_m = x_{F,0} \cdot (1 + R)$	$x_F = x_{F,0}$
Dilution (F/A)	$x_D = x_{D,0}$	$x_m = x_{D,0} \cdot (1 + \frac{1}{R})$	$x_O = 0.2095 \cdot (1 - x_F - x_{D,0})$
Dilution (D/F)	$x_F = x_{F,0} + x_{D,0} - x_D$	$x_m = x_{F,0} + x_{D,0} = C$	$x_O = 0.2095(1 - x_{F,0} - x_{D,0})$

(designated with a subscript 0). Analytical purge curves (listed in Table 6.1) are more complex than the diluting curves (listed in Table 6.2), since the impact of additive is shared by all constituents, while the background species is constant in diluting operations. In a ternary flammability diagram, the purge curves is much simpler, just draw a straight line from the initial point (sample point or initial point) to the extreme condition (100 % diluent point).

Again, we need to apply these purge curves to a sample problem to appreciate their utilities in guiding purge operations.

There is a static mixture sample taken in a confined space (type I problem), with 10 % methane, 20 % excess nitrogen and 70 % normal air. Perform the following analysis on each flammability diagrams and solve it with analytical solutions. Answer the following questions.

a. How to make the mixture safe (non-flammable) by adding fuel?
b. How to make the mixture safe (non-flammable, non-explosive or non-ignitable) by adding diluent?
c. How to make the mixture safe (fuel-lean) by adding air?

6.3.2 *Purging Operation in the Standard Flammability Diagram*

Purging operations in the standard flammability diagram are presented in Fig. 6.8.

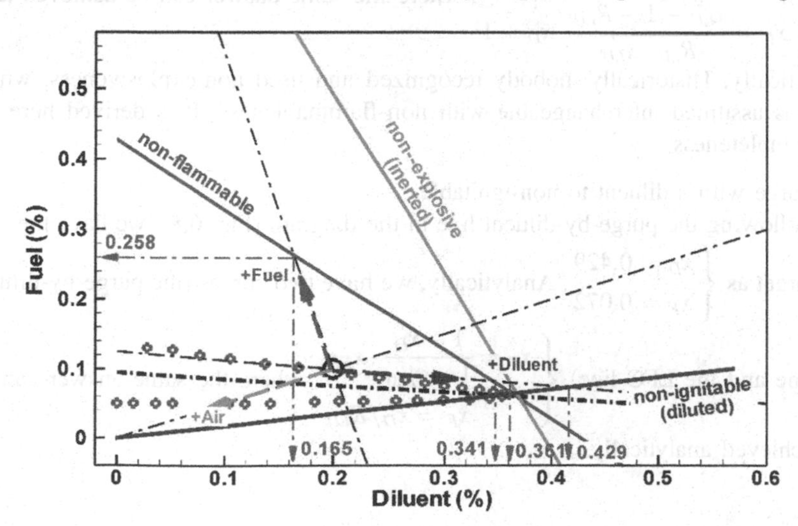

Fig. 6.8 Purging operations in a standard flammability diagram

a. Purge with a fuel to non-flammable.
Following the purge-by-fuel line in the diagram (Fig. 6.8), we have the safe target $\begin{cases} x_D = 0.165 \\ x_F = 0.258 \end{cases}$ for non-flammability. However, we can also solve the two coupling curves (the purge-by-fuel line and the MOC line) defined by $\begin{cases} x_D = \dfrac{1 - x_F}{1 - x_{D,0}} \cdot x_{D,0} \\ x_D + x_F = 1 - 4.773 \cdot \lambda_1 = 0.43 \end{cases}$, where the same answer can be achieved analytically.

b. Purge with a diluent to non-flammable.
Following the purge-by-diluent line in the diagram (Fig. 6.8), we have the safe target as $\begin{cases} x_D = 0.341 \\ x_F = 0.082 \end{cases}$. However, we can also solve the two coupling curves (the purge-by-diluent line and the MOC line) defined by $\begin{cases} x_F = \dfrac{1 - x_D}{1 - x_{D,0}} \cdot x_{F,0} \\ x_D + x_F = 1 - 4.773 \cdot \lambda_1 = 0.43 \end{cases}$, where the same answer can be achieved analytically.

c. Purge with a diluent to non-explosive
Following the purge-by-diluent line in the diagram (Fig. 6.8), we have the safe target as $\begin{cases} x_D = 0.361 \\ x_F = 0.080 \end{cases}$. However, we can also solve the two coupling curves (the purge-by-diluent line and the LOC line) defined by $\begin{cases} x_F = \dfrac{1 - x_D}{1 - x_{D,0}} \cdot x_{F,0} \\ x_F = \dfrac{x_{LU} - 1 - R_{LU}}{R_{LU} \cdot x_{LU}} \cdot x_D + 1 \end{cases}$, where the same answer can be achieved analytically. Historically, nobody recognized and used non-explosiveness, which was assumed interchangeable with non-flammableness. It is derived here for completeness.

d. Purge with a diluent to non-ignitable
Following the purge-by-diluent line in the diagram (Fig. 6.8), we have the safe target as $\begin{cases} x_D = 0.429 \\ x_F = 0.072 \end{cases}$. Analytically, we have two curves (the purge-by-diluent line and the LFC line) $\begin{cases} x_F = \dfrac{1 - x_D}{1 - x_{D,0}} \cdot x_{F,0} \\ x_F = x_D / R_{LU} \end{cases}$, where the same answer can be achieved analytically.

e. Purge with air will not go to non-explosive or non-ignitable, but to the fuel-lean state only, located at $\begin{cases} x_D = 0.10 \\ x_F = 0.05 \end{cases}$.

6.3.3 Purging Operations in the Diluted Flammability Diagram

Purging operations in the diluted flammability diagram are presented in Fig. 6.9.

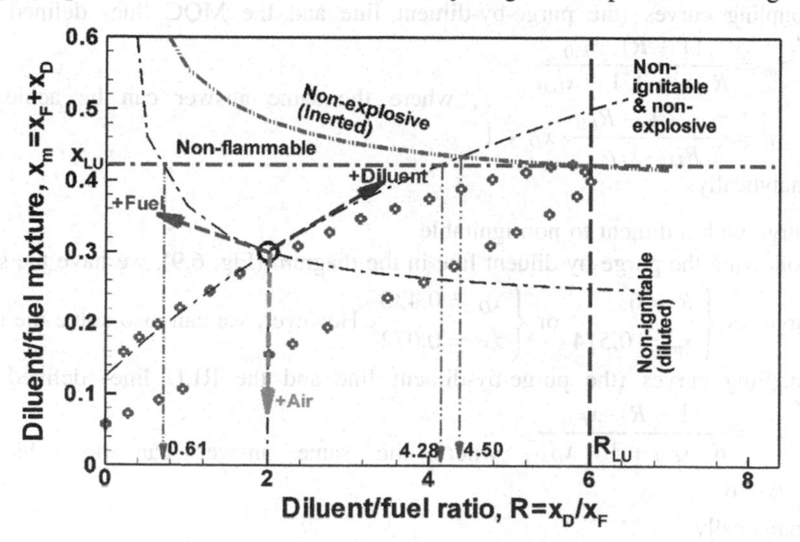

Fig. 6.9 Purging operations in the diluted flammability diagram

a. Purge with a fuel to non-flammable.
Following the purge-by-fuel line in the diagram (Fig. 6.9), we have the safe target as $\begin{cases} R = 0.61 \\ x_m = 0.43 \end{cases}$ or $\begin{cases} x_D = 0.163 \\ x_F = 0.267 \end{cases}$. However, we can also solve the two coupling curves (the purge-by-fuel line and the MMF line) defined by $\begin{cases} x_m = \dfrac{(1+R)\cdot x_{D,0}}{R\cdot(1-x_{F,0})+x_{D,0}} \\ x_m = x_D + x_F = 1 - 4.773\cdot\lambda_1 = 0.43 \end{cases}$, where the same answer can be achieved analytically.

b. Purge with a diluent to non-flammable
Following the purge-by-diluent line in the diagram (Fig. 6.9), we have the safe target as $\begin{cases} R = 4.28 \\ x_m = 0.43 \end{cases}$ or $\begin{cases} x_D = 0.348 \\ x_F = 0.082 \end{cases}$. However, we can also solve the two

coupling curves (the purge-by-diluent line and the MMF line) defined by $\begin{cases} x_m = \dfrac{(1+R)\cdot x_{F,0}}{R\cdot x_{F,0}+1-x_{D,0}} \\ x_m = x_D + x_F = 1 - 4.773\cdot\lambda_1 = 0.43 \end{cases}$, where the same answer can be achieved analytically.

c. Purge with a diluent to non-explosive
Following the purge-by-diluent line in the diagram (Fig. 6.9), we have the safe target as $\begin{cases} R = 4.50 \\ x_m = 0.44 \end{cases}$ or $\begin{cases} x_D = 0.361 \\ x_F = 0.080 \end{cases}$. However, we can also solve the two coupling curves (the purge-by-diluent line and the MOC line) defined by $\begin{cases} x_m = \dfrac{(1+R)\cdot x_{F,0}}{R\cdot x_{F,0}+1-x_{D,0}} \\ x_F = \dfrac{x_{LU}-1-R_{LU}}{R_{LU}\cdot x_{LU}}\cdot x_D + 1 \end{cases}$, where the same answer can be achieved analytically.

d. Purge with a diluent to non-ignitable
Following the purge-by-diluent line in the diagram (Fig. 6.9), we have the safe target as $\begin{cases} R = 6 \\ x_m = 0.514 \end{cases}$ or $\begin{cases} x_D = 0.430 \\ x_F = 0.072 \end{cases}$. However, we can also solve the two coupling curves (the purge-by-diluent line and the RLU line) defined by $\begin{cases} x_m = \dfrac{(1+R)\cdot x_{F,0}}{R\cdot x_{F,0}+1-x_{D,0}}, \\ R = 6 \end{cases}$ where the same answer can be achieved analytically.

e. Purge with air will not go to non-flammable or non-ignitable, but to a lean state located at $\begin{cases} R = 2 \\ x_m = 0.15 \end{cases}$ or $\begin{cases} x_D = 0.10 \\ x_F = 0.05 \end{cases}$.

6.3.4 Purging Operations in the Explosive Triangle Diagram

Purging operations in the explosive triangle diagram are presented in Fig. 6.10.

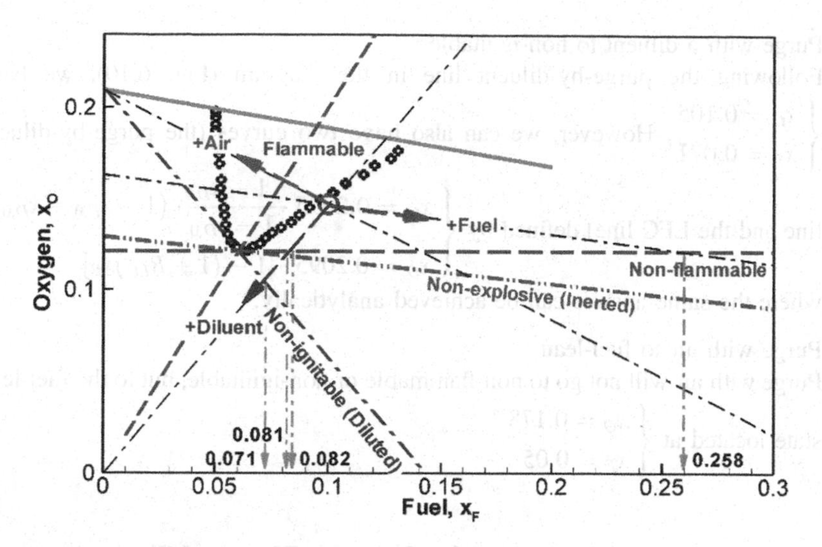

Fig. 6.10 Purging operations in the explosive triangle diagram

a. Purge with a fuel to non-flammable.
   Following the purge-by-fuel line in the diagram (Fig. 6.10), we have $\begin{pmatrix} x_O = 0.121 \\ x_F = 0.258 \end{pmatrix}$. However, we can also have two curves (the purge-by-fuel line and the MOC line) defined by $\begin{cases} x_O = 0.2095 \cdot \dfrac{1 - x_F}{1 - x_{F,0}} \cdot (1 - x_{F,0} - x_{D,0}) \\ x_O = \lambda_1 = 0.121 \end{cases}$, where the same answer can be achieved analytically.

b. Purge with a diluent to non-flammable
   Following the purge-by-diluent line in the diagram (Fig. 6.10), we have $\begin{cases} x_O = 0.121 \\ x_F = 0.082 \end{cases}$. However, we can also have two curves (the purge-by-diluent line and the MOC line) defined by $\begin{cases} x_O = 0.2095 \cdot \dfrac{1 - x_D}{1 - x_{D,0}} \cdot (1 - x_{F,0} - x_{D,0}) \\ x_O = \lambda_1 = 0.121 \end{cases}$, where the same answer can be achieved analytically.

c. Purge with a diluent to non-explosive
   Following the purge-by-diluent line in the diagram (Fig. 6.10), we have $\begin{cases} x_O = 0.116 \\ x_F = 0.081 \end{cases}$. However, we can also have two curves (the purge-by-diluent line and the LOC line) defined by $\begin{cases} x_O = 0.2095 \cdot \dfrac{1 - x_D}{1 - x_{D,0}} \cdot (1 - x_{F,0} - x_{D,0}) \\ x_O = \lambda_2 (1 - x_F) \end{cases}$, where the same answer can be achieved analytically.

d. Purge with a diluent to non-ignitable

Following the purge-by-diluent line in the diagram (Fig. 6.10), we have $\begin{cases} x_O = 0.105 \\ x_F = 0.071 \end{cases}$. However, we can also have two curves (the purge-by-diluent line and the LFC line) defined by $\begin{cases} x_O = 0.2095 \cdot \dfrac{1 - x_D}{1 - x_{D,0}} \cdot (1 - x_{F,0} - x_{D,0}) \\ x_O = 0.2095 \cdot [1 - (1 + R_{LU})x_F] \end{cases}$, where the same answer can be achieved analytically.

e. Purge with air to fuel-lean

Purge with air will not go to non-flammable or non-ignitable, but to the fuel-lean state located at $\begin{cases} x_O = 0.178 \\ x_F = 0.05 \end{cases}$.

6.3.5 Purging Operations in the Ternary Flammability Diagram

Purging operations in the ternary flammability diagram are presented in Fig. 6.11.

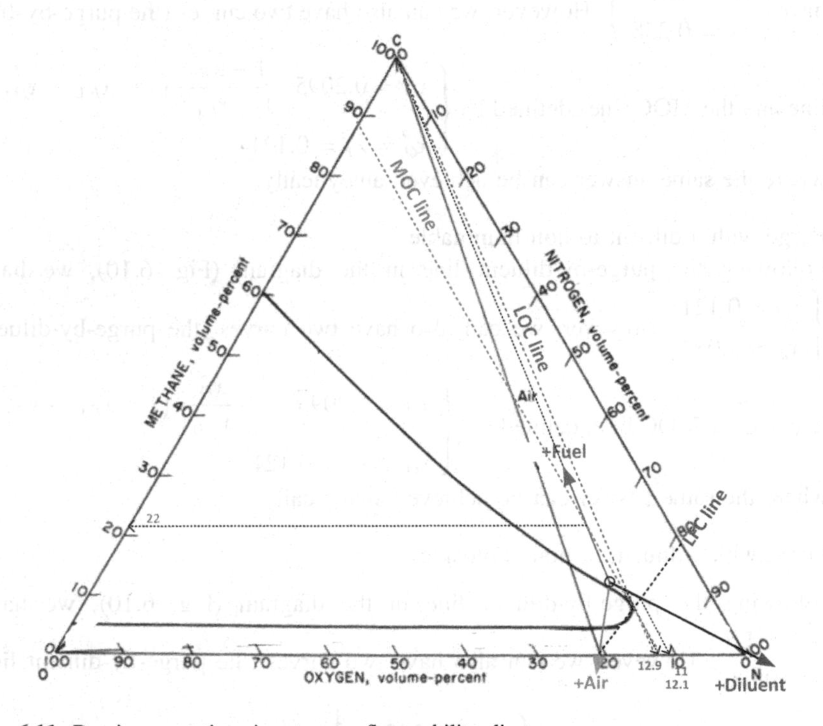

Fig. 6.11 Purging operations in a ternary flammability diagram

a. Using experimental data of methane, the flammability diagrams of methane are shown in Fig. 6.11. The sample point lies outside the flammable domain, but still above the MOC line, so it is conditional-flammable.

b. Purge with air to the fuel-lean state, the critical safe condition is $\begin{cases} x_O = 0.19 \\ x_F = 0.05 \end{cases}$, which is the cross point of the purge-by-air line and LFC line.

c. The purging-by-fuel line will cross the MOC line at $\begin{cases} x_O = 0.121 \\ x_F = 0.258 \end{cases}$.

d. Purging by diluent will cross the MOC/LOC/LFC lines respectively. The mixture will be non-explosive at $\begin{cases} x_O = 0.116 \\ x_F = 0.081 \end{cases}$, non-flammable at $\begin{cases} x_O = 0.121 \\ x_F = 0.082 \end{cases}$, and non-ignitable at $\begin{cases} x_O = 0.105 \\ x_F = 0.071 \end{cases}$.

e. Purge with air to fuel-lean
Purge with air will not go to non-flammable or non-ignitable, but to the fuel-lean state located at $\begin{cases} x_O = 0.178 \\ x_F = 0.05 \end{cases}$.

6.4 Dilution in Flammability Diagrams

Sometimes, the original mixture in a confined space does not have a third species, such as a full fuel tank or an empty fuel tank. The purge of the space is realized by introducing an agent to displace another. In other times, the fuel/diluent combination can be strictly controlled, such as a fuel stream. The concentration rise of one species is realized by decreasing another. In both cases, they are dilution problems, which have only two species floating with a third species (background air) constant.

If the mixture is coming from a dynamic stream of fuel/oxidizer/diluent, adjusting two concentrations will not affect the concentration of the third, it is treated as a diluting problem (as compared to the static mixture in a confinement, a purge problem). Here there are three reversible diluting scenarios, adding fuel by displacing air (designated as F/A), adding diluent by displacing air (designated as D/A), adding diluent by displacing fuel (designated as F/D). Here we can treat the dynamic diluting process by moving the operation point along one of the six directions, as shown in Figs. 6.12, 6.13, 6.14 and 6.15. Note, dilution means only two concentrations are adjustable, while the third concentration stays constant. So dilution (D/A) is in fact an iso-fuel line, dilution (F/A) is in fact an iso-diluent line, while dilution (F/D) is in fact an iso-air line. From these principles, we can derive the analytical operating curves in Table 6.2. It is difficult to express the analytical solutions in a ternary diagram.

Here is a typical sample problem for diluting operations.

A stream of gases with 10 % methane, 20 % nitrogen and 70 % air are released and mixed in air. Answer the following questions.

a. How to make the mixture non-flammable or non-ignitable by adding fuel?
b. How to make the mixture non-flammable or non-ignitable by adding diluent?
c. How to make the mixture non-flammable or non-ignitable by adding air?

6.4.1 Diluting Operations in a Standard Flammability Diagram

Diluting operations in the standard flammability diagram are presented in Fig. 6.12.

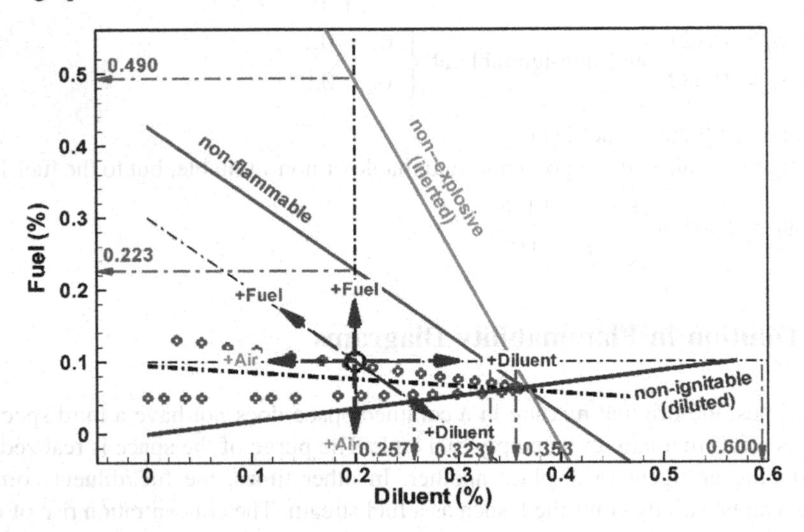

Fig. 6.12 Diluting operations in a standard flammability diagram

Solution:

a. Using experimental data of methane, the flammability diagrams of methane are shown in Fig. 6.12. The mixture point falls above the flammable zone, but below non-flammable line. It is still dangerous, since MOC is not guaranteed.

b. In order to inertize the fuel mixture, more fuel is added into the mixture by displacing the air or the diluent. Displacing air, the critical safe target is $\begin{cases} x_F = 0.223 \\ x_D = 0.200 \end{cases}$, which will reduce the oxygen centration to MOC = 0.121. Displacing diluent, the mixture goes to nowhere, still in the flammable zone.

c. In order to dilute the fuel mixture, more diluent is added into the mixture by displacing the air or displacing the fuel. Displacing the air, the critical diluent fraction is $\begin{cases} x_F = 0.100 \\ x_D = 0.323 \end{cases}$ for non-flammable, $\begin{cases} x_F = 0.100 \\ x_D = 0.363 \end{cases}$ for non-explosive, $\begin{cases} x_F = 0.100 \\ x_D = 0.600 \end{cases}$ for non-ignitable. Displacing the fuel, the critical diluent fraction is $\begin{cases} x_F = 0.043 \\ x_D = 0.257 \end{cases}$ 0.257 for non-ignitable.

d. In order to dilute the mixture, additional air is added into the mixture by displacing the diluent or displacing the fuel. Displacing the diluent, the mixture goes to nowhere. Displacing the fuel, additional air can decrease the fuel concentration to $\begin{cases} x_F = 0.05 \\ x_D = 0.200 \end{cases}$ for the fuel-lean state or to the non-ignitable state of $\begin{cases} x_F = 0.033 \\ x_D = 0.200 \end{cases}$. However, the flammable zone will be crossed. This is not the recommended safe operation while crossing the flammable zone.

6.4.2 Diluting Operations in a Diluted Flammability Diagram

Diluting operations in the diluted flammability diagram are presented in Fig. 6.13.

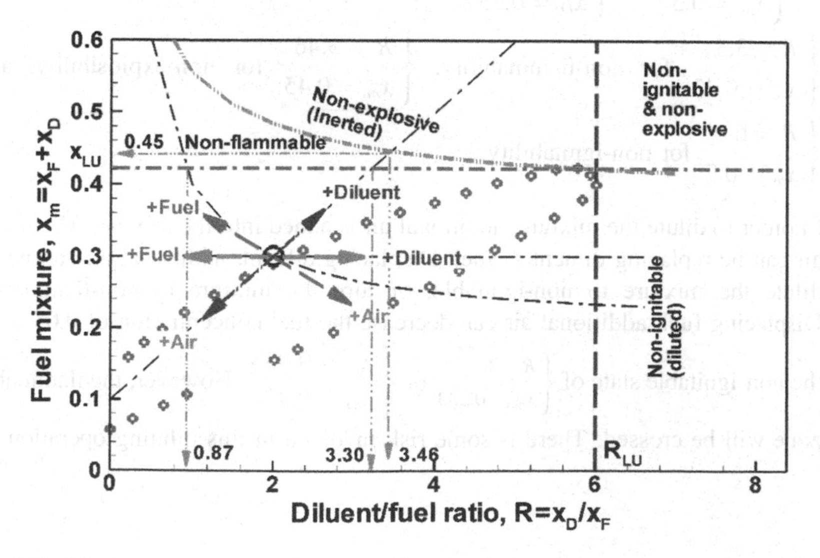

Fig. 6.13 Diluting operations in a Diluted flammability diagram

Solution:

a. Using experimental data of methane, the flammability diagrams of methane are shown in Fig. 6.13. The mixture point falls above the flammable zone, but still under the non-flammable zone, or in the fuel-rich zone. It is still dangerous, since MOC is not guaranteed.

b. In order to inertize the fuel mixture, more fuel is added into the mixture by displacing the air. The critical target for non-flammability is found to be $\begin{cases} R = 0.87 \\ x_m = 0.43 \end{cases}$ by solving $\begin{cases} x_m = \left(1 + \frac{1}{R}\right) \cdot x_{D,0} \\ x_m = x_{LU} = 0.43 \end{cases}$. That means $\begin{cases} x_F = 0.2 \\ x_D = 0.23 \end{cases}$ (by solving $\begin{cases} x_F + x_D = x_m = 0.43 \\ x_D / x_F = 0.87 \end{cases}$), which will reduce the oxygen centration to MOC = 0.121. The critical target for non-explosibility is $\begin{cases} R = 0.417 \\ x_m = 0.68 \end{cases}$ by solving $\begin{cases} x_m = \left(1 + \frac{1}{R}\right) \cdot x_{D,0} \\ x_m = \dfrac{(1 - 4.773\lambda_2) \cdot (1 + R)}{1 - 4.773\lambda_2 + R} \end{cases}$. So this mixture can be inertised to non-flammable or non-explosive by F/A dilution.

c. In order to dilute the fuel mixture, more diluent is added into the mixture by displacing the air or by displacing the fuel. Displacing the fuel, the critical target is $\begin{cases} R = 6 \\ x_m = 0.3 \end{cases}$ or $\begin{cases} x_F = 0.043 \\ x_D = 0.257 \end{cases}$. Displacing the air, the critical target is $\begin{cases} R = 3.3 \\ x_m = 0.43 \end{cases}$ for non-flammability, $\begin{cases} R = 3.46 \\ x_m = 0.45 \end{cases}$ for non-explosibility, and $\begin{cases} R = 6 \\ x_m = 0.7 \end{cases}$ for non-ignitability.

d. In order to dilute the mixture, additional air is added into the mixture. However, air can be replacing diluent or fuel. Displacing diluent, additional air can never dilute the mixture to non-ignitable, or turn the mixture to non-flammable. Displacing fuel, additional air can decrease the fuel concentration of 0.05 or to the non-ignitable state of $\begin{cases} R = 6 \\ x_m = 0.233 \end{cases}$ or $\begin{cases} x_F = 0.033 \\ x_D = 0.2 \end{cases}$. However, the flammable zone will be crossed. There is some risk involved in this diluting operation.

6.4.3 Diluting Operations in an Explosive Triangle Diagram

Diluting operations in the explosive triangle diagram are presented in Fig. 6.14.

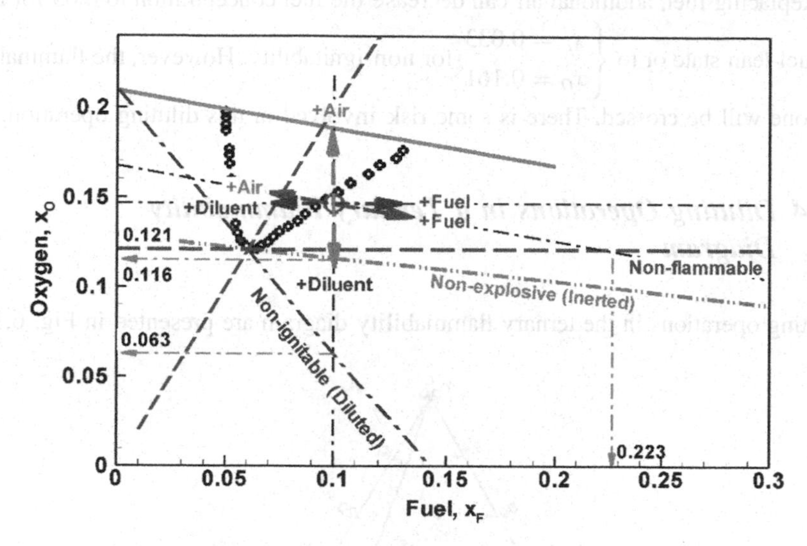

Fig. 6.14 Diluting operations in an explosive triangle diagram

Solution:

a. Using experimental data of methane, the flammability diagrams of methane are shown in Fig. 6.14. The mixture point falls outside the flammable zone, but still above the non-flammable zone, or in the fuel-rich zone. It is still dangerous, since MOC is not reached.

b. In order to decrease the oxygen level to MOC, additional fuel is used for dilution. Displacing air, the safe target is $\begin{cases} x_F = 0.223 \\ x_O = 0.121 \end{cases}$ for non-flammability and $\begin{cases} x_F = 0.480 \\ x_O = 0.067 \end{cases}$ for non-explosibility. Displacing diluent, there is no safe target available.

c. Diluent can be used for displacing fuel or air. Displacing fuel, the critical safe target is $\begin{cases} x_F = 0.043 \\ x_O = 0.147 \end{cases}$ for non-ignitability. Displacing air, the critical safe target is $\begin{cases} x_F = 0.100 \\ x_O = 0.121 \end{cases}$ for non-flammability, $\begin{cases} x_F = 0.100 \\ x_O = 0.116 \end{cases}$ for non-explosiveness, and $\begin{cases} x_F = 0.100 \\ x_O = 0.063 \end{cases}$ for non-ignitability.

d. In order to dilute the mixture, additional air is added into the mixture. However, air can be replacing diluent or fuel. Replacing diluent, additional air can never dilute the mixture to non-ignitable, or inert the mixture to non-flammable. Replacing fuel, additional air can decrease the fuel concentration to 0.05 for the fuel-lean state or to $\begin{cases} x_F = 0.033 \\ x_O = 0.161 \end{cases}$ for non-ignitability. However, the flammable zone will be crossed. There is some risk involved in this diluting operation.

6.4.4 Diluting Operations in a Ternary Flammability Diagram

Diluting operations in the ternary flammability diagram are presented in Fig. 6.15.

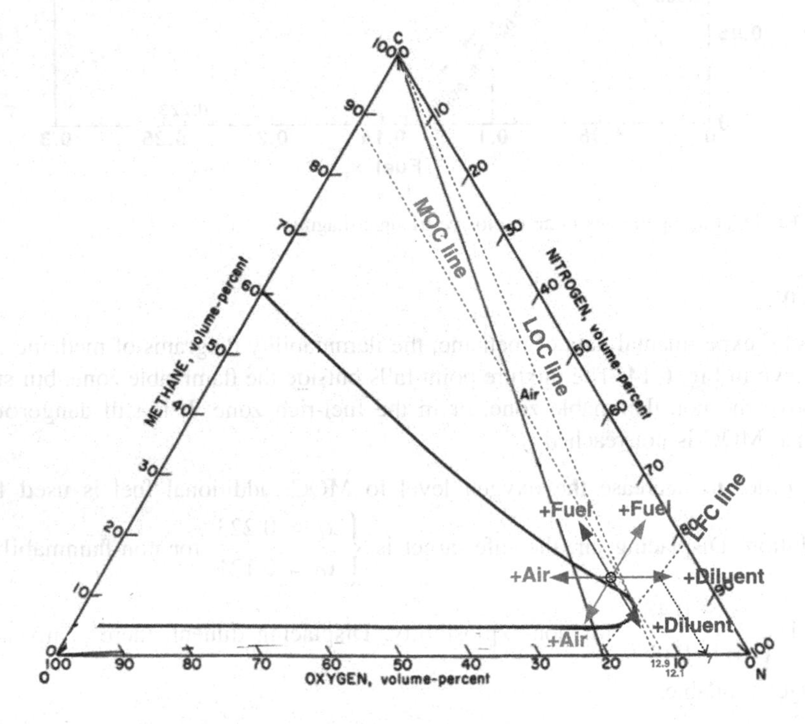

Fig. 6.15 Diluting operations in a ternary flammability diagram

Solution:

a. Using experimental data of methane, the flammability diagrams of methane are shown in Fig. 6.15. The mixture point falls above the flammable zone, but below the non-flammable zone, or in the fuel-rich zone. It is still dangerous, since the condition of MOC is not satisfied.

b. In order to dilute oxygen (or inertise the fuel mixture), more fuel is added into the mixture by displacing air. The critical condition is the cross point between the displacing-by-air fuel line and the MOC line, or $\begin{cases} x_O = 0.121 \\ x_F = 0.11 \end{cases}$, which will reduce the oxygen centration to MOC = 0.12. So the mixture can be inertised by adding fuel.

c. In order to dilute the fuel mixture, more diluent is added into the mixture by displacing the air or displacing fuel. For displacing air, the critical condition is $\begin{cases} x_O = 0.121 \\ x_F = 0.10 \end{cases}$ for non-flammability, $\begin{cases} x_O = 0.116 \\ x_F = 0.10 \end{cases}$ for non-explosibility and $\begin{cases} x_O = 0.07 \\ x_F = 0.10 \end{cases}$ for non-ignitability. For displacing fuel, the critical condition is $\begin{cases} x_O = 0.147 \\ x_F = 0.07 \end{cases}$ for non-ignitability.

d. In order to dilute the mixture, additional air is added into the mixture. However, air can be replacing diluent or fuel. Displacing diluent, additional air can never dilute the mixture to non-ignitable, or inter the mixture to non-flammable. Displacing fuel, additional air can decrease the fuel concentration to 0.05 for the fuel-lean state. However, the flammable zone will be crossed. There is some risk involved in this diluting operation.

6.5 Summary on Purge and Dilution in Diagrams

From the above sample mixture, we have 11 possibilities (7 diluting plus 4 purge operations) to reach a safe target. By introducing and differentiating, the concepts of non-explosive, non-flammable and non-ignitable, flammable domains are better presented with analytical boundaries. All these boundaries are developed from their physical meanings. A comparison of results of above operations is shown in Table 6.3. It is found that all flammability diagrams are equivalent to each other.

With the analytical solutions, the diluting and purging operations are differentiated and presented in the flammability diagrams. This helps to turn the old state diagram into the new operation-specific process diagrams. Thus, these diagrams can provide more information on guiding the safe handling of flammable mixtures.

Since all diagrams are compared here, a natural question arises, which one is best? From the above analysis, we can see that all diagrams are equivalent in terms of conservation of energy and volume. However, from the utility perspective, we can have the following rank, diluted > triangle > standard > ternary. Coward's diluted flammability diagram is best by providing rectangular flammable boundary and critical parameters (MMR/MMF), while the diluent/fuel ratio is a more preferred operation parameter in industry. Therefore, it is most useful in presenting data.

Table 6.3 Target safe operation points via various operations

Operation	Standard	Diluted	Triangle	Ternary
Dilution (D/A) to non-flammable	$x_F = 0.10$ $x_D = 0.323$	$x_m = 0.43$ $R = 3.3$	$x_F = 0.10$ $x_O = 0.121$	$x_F = 0.10$ $x_O = 0.121$
Dilution (D/A) to non-explosive	$x_F = 0.10$ $x_D = 0.363$	$x_m = 0.43$ $R = 3.46$	$x_F = 0.10$ $x_O = 0.116$	$x_F = 0.10$ $x_O = 0.116$
Dilution (D/A) to non-ignitable	$x_F = 0.10$ $x_D = 0.60$	$x_m = 0.70$ $R = 6$	$x_F = 0.10$ $x_O = 0.063$	$x_F = 0.10$ $x_O = 0.063$
Dilution (F/A) to non-flammable	$x_F = 0.223$ $x_D = 0.20$	$x_m = 0.43$ $R = 0.87$	$x_F = 0.223$ $x_O = 0.121$	$x_F = 0.223$ $x_O = 0.121$
Dilution (F/A) to non-explosive	$x_F = 0.49$ $x_D = 0.20$	$x_m = 0.68$ $R = 0.417$	$x_F = 0.480$ $x_O = 0.067$	$x_F = 0.480$ $x_O = 0.067$
Dilution (D/F) to non-flammable	$x_F = 0.043$ $x_D = 0.257$	$x_m = 0.43$ $R = 6$	$x_F = 0.043$ $x_O = 0.147$	$x_F = 0.043$ $x_O = 0.147$
Dilution (A/D) to non-flammable	$x_F = 0.033$ $x_D = 0.20$	$x_m = 0.233$ $R = 6$	$x_F = 0.033$ $x_O = 0.161$	$x_F = 0.033$ $x_O = 0.161$
Purge (D) to non-flammable	$x_F = 0.082$ $x_D = 0.341$	$x_m = 0.43$ $R = 4.28$	$x_F = 0.082$ $x_O = 0.121$	$x_F = 0.082$ $x_O = 0.121$
Purge (D) to non-explosive	$x_F = 0.081$ $x_D = 0.361$	$x_m = 0.43$ $R = 4.50$	$x_F = 0.080$ $x_O = 0.116$	$x_F = 0.080$ $x_O = 0.116$
Purge (D) to non-ignitable	$x_F = 0.072$ $x_D = 0.429$	$x_m = 0.514$ $R = 6$	$x_F = 0.072$ $x_O = 0.105$	$x_F = 0.072$ $x_O = 0.105$
Purge (F) to non-flammable	$x_F = 0.258$ $x_D = 0.165$	$x_m = 0.43$ $R = 0.61$	$x_F = 0.258$ $x_O = 0.121$	$x_F = 0.258$ $x_O = 0.121$

Explosive triangle is a second best, since LOC and LFC are clearly defined, and occupy a significant portion in the domain. So there is sufficient resolution in reading data. The standard diagram has a small area for operations, so data are difficult to read. Ternary diagram has an even smaller area for dilution and purge. The boundaries do not have analytical expressions. In addition, experimental data to display in a ternary diagram need special conversion, further reducing its attractiveness for safe operations. Simply by comparing the effective readable areas, ternary diagram is inferior to other diagrams.

6.6 Application on Tank Operations

Tank fire is a serious problem in the petrochemical industry [115]. One application of the flammability diagram is to guide the filling-tank operations on inerting a flammable mixture within a confined space. Filling is a typical type I problem for

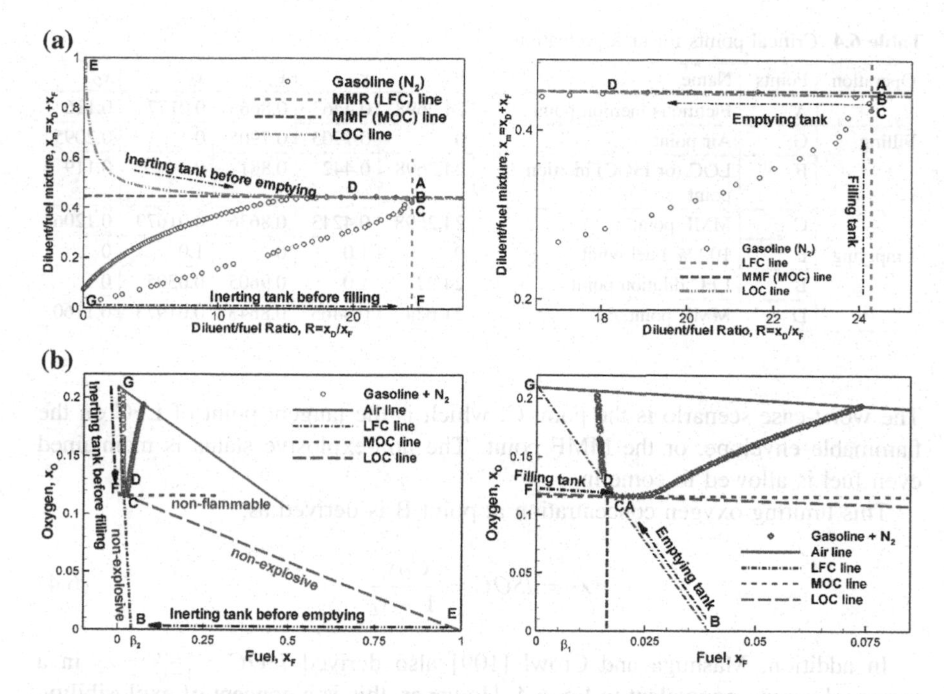

Fig. 6.16 Flammability diagrams for filling and emptying operations. **a** Diluted flammability diagram and local zoom up. (*Filling* GFC, *Emptying* EBD), **b** Explosive triangle diagram and local zoom up (*Filling* GFC, *Emptying* EBD)

controlling the oxygen, and emptying is a typical type II problem for controlling the fuel. Therefore, we have two types of problem to solve in a fuel tank (or a confined space).

In order to determine the diluent requirement for inerting a tank prior to filling the tank with gasoline, we have a reduced flammability diagram and an explosive triangle shown in Fig. 6.16. Critical points in Fig. 6.16 are listed in Table 6.4.

6.6.1 Safe Operations Before Filling a Liquid Tank (Type I Problem)

Before filling an empty fuel tank, the air has to be fully inertised before the filling starts. That means the oxygen level should be controlled. Here the inertion process started from the air point (G), to the inertion point (F). The inertion point F has an oxygen concentration of LOC or ISOC (in-service oxygen concentration). The ISOC represents the maximum oxygen concentration that just avoids the flammability zone. Then stop adding diluent, and allow the fuel to fill the tank. Once the fuel evaporates, its vapor will move the point along the LOC line, from F to C.

Table 6.4 Critical points for tank protections

Operation	Points	Name	R	x_m	x_D	x_F	x_O
	A	Fictitious inertion point	24.2698	0.4463	0.8663	0.0177	0.116
Filling	G	Air point	0	0.7905	0.7905	0	0.2095
	F	LOC (or ISOC) inertion point	24.2698	0.442	0.881	0	0.119
	C	MMF point	24.2698	0.4243	0.8636	0.01679	0.1206
Emptying	E	100 % Fuel point	0	1.0	0	1.0	0
	B	LFC dilution point	24.27	1.0	0.9605	0.0395	0
	D	MMR point	21.624	0.4463	0.8643	0.01973	0.1160

The worst-case scenario is the point C, which is the tangent point of LFC on the flammable envelope, or the MMF point. The non-explosive status is maintained even fuel is allowed to come in.

This limiting oxygen concentration at point B is derived as

$$\lambda_2 = ISOC = \frac{C_O x_L}{1 - x_L} \tag{6.4}$$

In addition, Mashuga and Crowl [109] also derived $ISOC = \frac{C_O x_L}{1-x_L} = \lambda_2$ in a ternary diagram, equivalent to Eq. 6.4. However, this is a concept of explosibility, which makes the mixture non-explosive, or more precisely, makes the flammable envelope shrink to nil.

Figure 6.16b shows the operating routes for protecting gasoline tanks by nitrogen in an explosive triangle diagram. It is a type I problem, but since the fuel is missing, the purge process is equivalent to a diluting operation. In order to fully inertize the air in the tank, an oxygen level of $\lambda_2 = 0.118$ will be reached by dilution to keep the mixture non-explosive at point B. After introducing the fuel into the tank, the oxygen concentration will drop to $\lambda_1 = 0.116$ (Point D) or further, still providing a non-explosive environment against flame propagation or explosive burning.

6.6.2 *Safe Operations Before Emptying a Liquid Tank (Type II Problem)*

Before emptying a fuel tank, the air space should be inerted before allowing more air to leak in. From the diluted flammable envelope (Fig. 6.16), the inertion process started from the fuel point E, and move to the (fuel) dilution point B. Once the fuel space is fully inerted at B, the emptying process can start, which allows more air come in to replace fuel. Under the inertion line LFC, the mixture is kept non-ignitable. The emptying process will increase the oxygen concentration along the LOC line, further to the point D. Point D is the cross point of LOC and MOC lines, or the MMR point, so the non-ignitable status is maintained even air is allowed to come in.

The (fuel) inertion point B is controlled by LFC, derived as

$$LFC = \beta_2 = OSFC = \frac{1}{1 + R_{LU}} \tag{6.5}$$

where the initial fuel concentration is $x_C = 1$ for a full tank space with fuel vapor. Note, Mashuga and Crowl [109] derived

$$OSFC = \frac{x_L}{1 - C_O x_L/\lambda} = \frac{\lambda_1}{C_O(1 - \lambda_1/\lambda)} = \frac{C_O \cdot x_L}{C_O \cdot x_{LU}} \tag{6.6}$$

where $\lambda_1 = C_O x_L$ is used in their derivation. Note for nitrogen in air, $Q_D = Q_N = 0.992 \approx 1$, so $x_F = x_L$. That is

$$OSFC = \frac{x_L}{x_{LU}} = \frac{x_F}{x_{LU}} = \frac{x_F}{x_D + x_F} = \frac{1}{1 + R_{LU}} \tag{6.7}$$

Obviously, their OSFC is equivalent to Eq. 6.5, if nitrogen is used as the diluent.

Graphically, the fuel vapor space will be purged (or diluted, since there is no air available yet) with a diluent (nitrogen), moving from point E to B. Then the emptying process will bring in air. However, since the operation points are all along or under the LFC line, the mixture is kept non-ignitable all the time. The worst-case scenario is the point D, on both the LFC line and the envelope. Here the vapor space is fully diluted (non-ignitable) before it has access to the air.

Figure 6.16b shows the operating routes for protecting gasoline tanks by nitrogen in an explosive triangle diagram. The fuel concentration should be dropped from one to β_2= 0.0398 (point F) by nitrogen dilution to keep the fuel stream non-ignitable in air. After introducing air into the tank, the fuel concentration will further drop to β_1= 0.0116 (point D) or further, still providing a non-flammable environment against flame propagation.

In summary, the purge process prior to a filling process is to control the oxygen, so the space is non-explosive before the filling starts. The emptying process is to control the fuel vapor level, so that the mixture is non-ignitable before the emptying starts. This result is consistent with the result from a ternary diagram [109]. In conclusion, this theory extends the diluted diagram-based operations by Planas-Cuchi et al. [116], and consistent with and equivalent to the theoretical framework set by Mashuga and Crowl [109] based on a ternary diagram. Both can be better presented in an explosive triangle diagram. Critical points are explained and unified satisfyingly using the concept of explosibility and ignitability in this work.

6.7 Problems and Solutions

6.7.1 *Flammable State in a Ternary Diagram*

Problem 6.1 If there is a methane mixture (50 % methane, 40 % oxygen and 10 % nitrogen) in a compartment of 1 m^3, how to purge the compartment with nitrogen, so the compartment is

a. Barely non-flammable?
b. Non-flammable?
c. Non-explosive?
d. Non-ignitable if the fuel mixture is released into air?

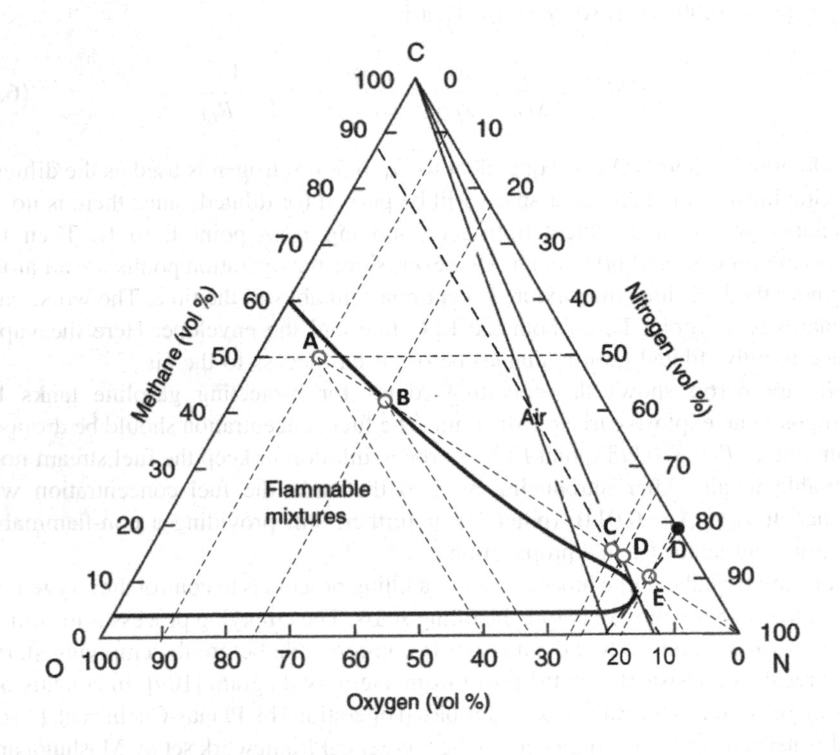

Solution:
First, plot the sample condition in the ternary diagram as point A. Draw a line connecting the 100 % Nitrogen point, which is the purge by nitrogen line. Then we can identify 4 points, which are solutions to the above questions.

a. The purge line crosses the flammable envelope at point B, which is read as 33 % Oxygen, 43 % Methane and 24 % Nitrogen. Therefore, the diluent requirement is $V_N = -V_O \cdot \ln\left(\frac{x_F}{x_{F,0}}\right) = -1 \cdot \ln\left(\frac{43\,\%}{50\,\%}\right) = 0.15\,\text{m}^3$.
b. The purge line crosses the MOC line at point C, which is read as 16 % Methane, 12 % Oxygen, and 72 % Nitrogen. Therefore, the diluent requirement is $V_N = -V_O \cdot \ln\left(\frac{x_F}{x_{F,0}}\right) = -1 \cdot \ln\left(\frac{16\,\%}{50\,\%}\right) = 1.14\,\text{m}^3$.
c. The purge line crosses the MOC line at point D, which is read as 15.5 % Methane, 11.5 % Oxygen, and 73 % Nitrogen. Therefore, the diluent requirement is $V_N = -V_O \cdot \ln\left(\frac{x_F}{x_{F,0}}\right) = -1 \cdot \ln\left(\frac{155\,\%}{50\,\%}\right) = 1.17\,\text{m}^3$.
d. To dilute the fuel mixture to non-ignitable, the safe target is point E, which is read as 10 % Methane, 7 % O_2 and 83 % Nitrogen. Since this is a dilution problem, the diluent requirement is $\frac{V_N + 1 \cdot 40\,\%}{1 + V_N} = 83\,\%$ or $V_N = 2.53\,\text{m}^3$.

6.7.2 Dilution and Purge for Gasoline Safety

Problem 6.2 If a sample taken from a confined crawl space underground shows the following composition: Gasoline 1.875 %, Oxygen 14.665 %, and Nitrogen 83.46 %. Answer the following questions using the flammability diagram taken from [4]

a. What is the flammable state of the mixture?
b. How to change the composition of the mixture to safety through dilution?
c. How to change the composition of the mixture to safety though purge?

Solution A: Using a diluted flammability diagram.

a. Take the experimental data from the standard flammability diagram, render the data in the diluted diagram. The critical parameters for safety is $R_{LU} = 24.27$, $x_{LU} = 0.4463$ for nitrogen envelope (Fig. 6.25). MOC is estimated as $\lambda_1 = 0.2095 \cdot (1 - x_{LU}) = 0.116$.
b. Since $x_o = 0.14665$, we have $x_D = 1 - x_F - 4.773 \cdot x_o = 0.2813$, so $R = 0.2813/0.01875 = 15$. Plot the sample point in diluted flammability diagram (Fig. 6.17). Obviously, it is flammable.

For a dilution operation, we have three possibilities controlled by different dilution curves.
For a dilution by diluent/air, the dilution curve is $x_m = x_{F,0} \cdot (1 + R)$.
For a dilution by fuel/air, the dilution curve is $x_m = x_{D,0} \cdot \left(1 + \frac{1}{R}\right)$.
For a dilution by diluent/fuel, the dilution curve is $x_m = x_{F,0} + x_{D,0} = C$.
All three curves are plotted on top of the flammable envelope. From these operations, we can make the following conclusions.

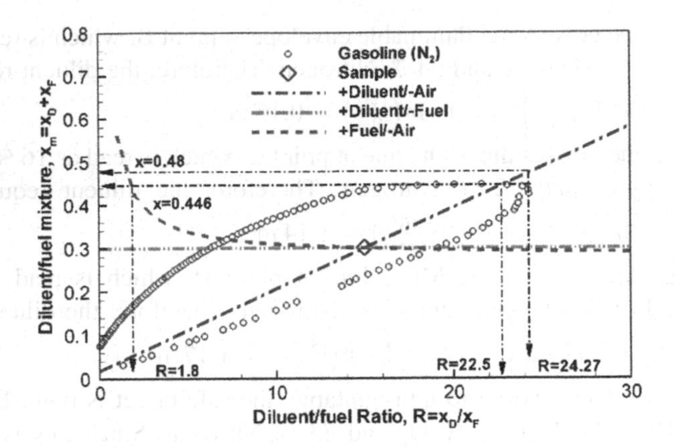

Fig. 6.17 Dilution operations in a diluted flammability diagram

(a) For a dilution by diluent/air, adding nitrogen will reach (R = 22.5, x = 0.446) for the non-flammable state, and (R = 24.27, x = 0.48) for the non-ignitable state.

(b) For a dilution by fuel/air, adding fuel will reach (R = 1.8, x = 0.446) for the non-flammable state, while adding air will reach (R = 24.27, x = 0.29) for non-ignitable state.

(c) For a dilution by diluent/fuel, adding diluent will reach (R = 24.27, x = 0.3) for the non-flammable state.

(d) Since $x_o = 0.14665$, we have $x_D = 1 - x_F - 4.773 \cdot x_o = 0.2813$, so $R = 0.2813/0.01875 = 15$. Plot the sample point in diluted flammability diagram (Fig. 6.18). Again, this mixture is flammable.

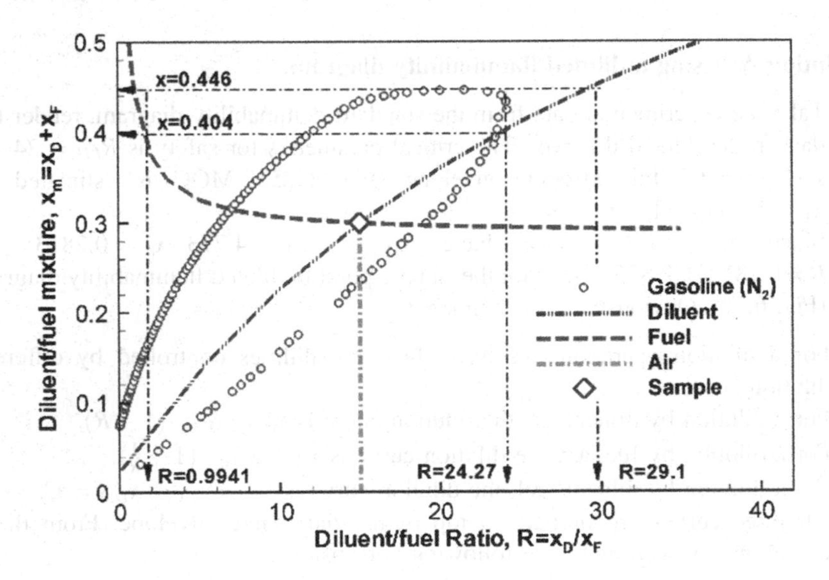

Fig. 6.18 Purge operations in a diluted flammability diagram

For a purge operation, we have three possibilities controlled by different purge curves.

For a purge operation by adding diluent, the purge curve is $x_m = \frac{x_{F,0} \cdot (1+R)}{R \cdot x_{F,0} + 1 - x_{D,0}}$.

For a purge operation by adding fuel, the purge curve is $x_m = \frac{x_{D,0} \cdot (1+R)}{R(1 - x_{F,0}) + x_{D,0}}$.

For a purge operation by adding air, the purge curve is $R = R_0$.

All three curves are plotted on the flammable envelope in Fig. 6.18. From these operations, we can make the following conclusions.

(a) For a purge by adding diluent, the sample will reach (R = 29.1, x = 0.4463) for the non-flammable state, and (R = 24.27, x = 0.404) for the non-ignitable state. They can be calculated directly from analytical curves $\begin{cases} x_m = 0.4463 \\ x_m = \dfrac{0.01875 \cdot (1+R)}{R \cdot 0.01875 + 1 - 0.28125} \end{cases}$ and $\begin{cases} R = 24.27 \\ x_m = \dfrac{0.01875 \cdot (1+R)}{R \cdot 0.01875 + 1 - 0.28125} \end{cases}$.

(b) For a dilution on fuel/air, adding fuel will reach (R = 0.9941, x = 0.4463) for the non-flammable state. It can be solved directly from $\begin{cases} x_m = 0.4463 \\ x_m = \dfrac{0.28125 \cdot (1+R)}{R \cdot (1 - 0.01875) + 0.28125} \end{cases}$.

(c) For a purge with air, it will never reach the non-ignitable state.

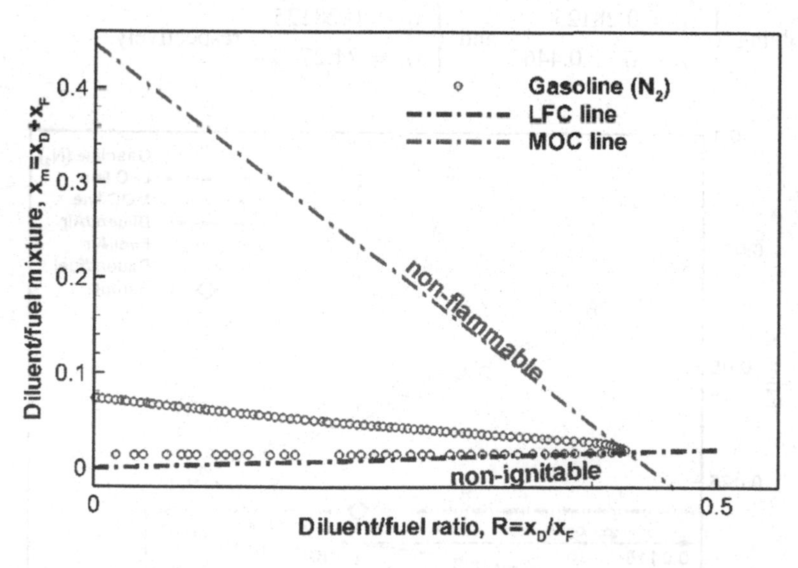

Fig. 6.19 Flammable state in a standard flammability diagram

Solution B: Using a standard flammability diagram.

a. Using $x_F = x_D/R_{LU}$ for drawing the dilution line, $x_F = x_{LU} - x_D$ for drawing the inertion line, we have the safe operation zones shown in Fig. 6.19.

b. Since $x_F = 0.01875$, we have $x_D = 1 - x_F - 4.773 \cdot x_O = 0.28125$. Plot the sample point in standard flammability diagram (Fig. 6.20).
For a dilution operation, we have three possibilities controlled by different dilution curves.
For a dilution on diluent/air, the dilution curve is $x_F = x_{F,0}$.
For a dilution on fuel/air, the dilution curve is $x_D = x_{D,0}$.
For a dilution on diluent/fuel, the dilution curve is $x_F = x_{F,0} + x_{D,0} - x_D$.
All three curves are plotted on the flammable envelope in Fig. 6.20. From these operations, we can make the following observations.

a. For a dilution on diluent/air, adding nitrogen will reach ($x_D = 0.455$, $x_F = 0.01875$) for the non-flammable state, and ($x_D = 0.4275$, $x_F = 0.01875$) for the non-ignitable state. They can be derived analytically by solving $\begin{cases} x_F = 0.01875 \\ x_D + x_F = 0.4463 \end{cases}$ and $\begin{cases} x_F = 0.01875 \\ x_D = 24.27 \cdot x_F \end{cases}$ respectively.

b. For a dilution on fuel/air, adding fuel will reach ($x_D = 0.28125$, $x_F = 0.165$) for the non-flammable state, while adding air will reach ($x_D = 0.28125$, $x_F = 0.01159$) for non-ignitable state. They can be derived analytically by solving $\begin{cases} x_D = 0.28125 \\ x_D + x_F = 0.4463 \end{cases}$ and $\begin{cases} x_D = 0.28125 \\ x_D = 24.27 \cdot x_F \end{cases}$ respectively.

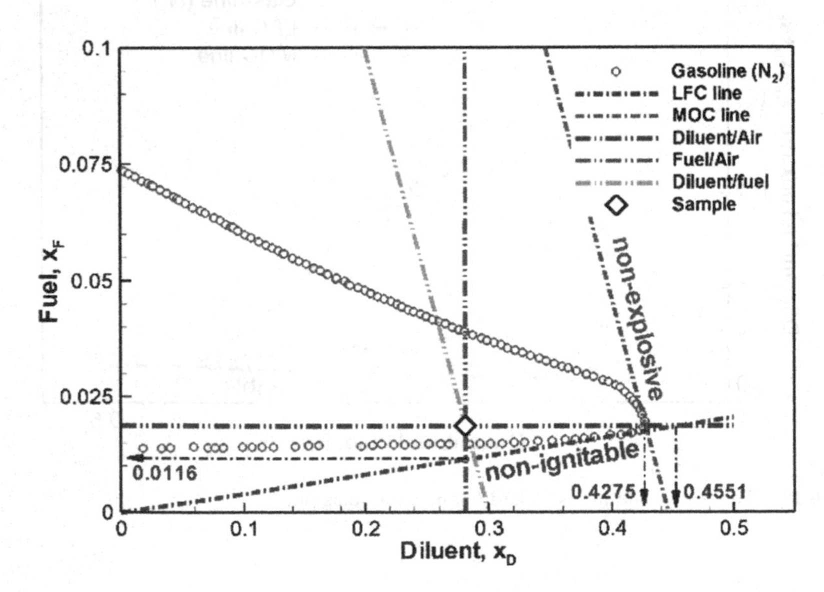

Fig. 6.20 Dilution operations in a standard flammability diagram

c. For a dilution on diluent/fuel, adding diluent will reach ($x_D = 0.28813$, $x_F = 0.01187$) for the non-flammable state. It can also be determined by solving $\begin{cases} x_D + x_F = 0.3 \\ x_D = 24.27 \cdot x_F \end{cases}$.

The dilution operations can be realized in a standard flammability diagram in Fig. 6.20.

d. Since $x_F = 0.01875$, we have $x_D = 1 - x_F - 4.773 \cdot x_O = 0.28125$. Plot the sample point in standard flammability diagram (Fig. 6.21).

For a purge operation, we have three possibilities controlled by different dilution curves.

For a purge operation by diluent, the purge curve is $x_F = \frac{1-x_D}{1-x_{D,0}} \cdot x_{F,0}$.

For a purge operation by fuel, the purge curve is $x_F = 1 - \frac{1-x_{F,0}}{x_{D,0}} \cdot x_D$.

For a purge operation by air, the purge curve is $x_F = x_D \cdot x_{F,0}/x_{D,0}$.

All three curves are plotted on the flammable envelope. From these operations, we can make the following conclusions.

a. For a purge by adding diluent, the operation point will reach ($x_D = 0.4583$, $x_F = 0.00918$) for the non-flammable state, and ($x_D = 0.3877$, $x_F = 0.0160$) for the non-ignitable state. They can be derived analytically by solving $\begin{cases} x_F = 0.4463 - x_D \\ x_F = \dfrac{1 - x_D}{1 - 0.28125} \cdot 0.01875 \end{cases}$ and $\begin{cases} x_F = x_D/24.27 \\ x_F = \dfrac{1 - x_D}{1 - 0.28125} \cdot 0.01875 \end{cases}$ respectively.

b. For a purge with additional fuel, the operation point will reach ($x_D = 0.2225$, $x_F = 0.2238$) for the non-flammable state, while it can never reach non-ignitable state by adding fuel. This can be derived analytically by solving $\begin{cases} x_F = 0.4463 - x_D \\ x_F = 1 - \dfrac{1 - 0.01875}{0.28125} \cdot x_D \end{cases}$.

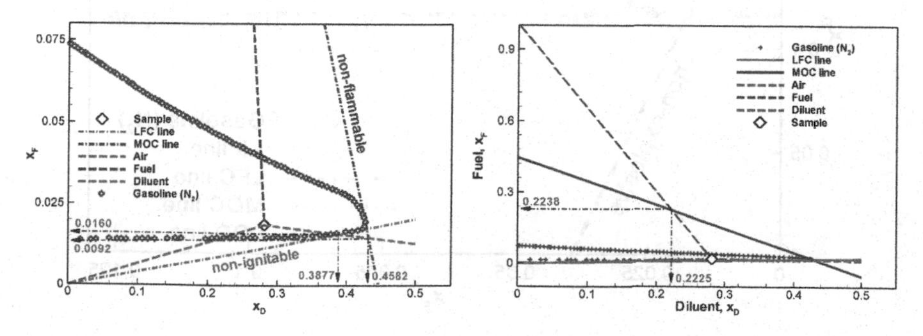

Fig. 6.21 Purge operations in a standard flammability diagram

c. For a purge with air, adding air will reach ($x_D = 0.28813, x_F = 0.01187$) for the non-flammable state. It can also be determined by solving $\begin{cases} x_D + x_F = 0.3 \\ x_D = 24.27 \cdot x_F \end{cases}$

The above purge operations can be realized in a standard flammability diagram shown in Fig. 6.21.

Solution C: using an Explosive Triangle diagram.

a. Using $x_O = 0.2095 - 0.2095 \cdot x_F$ for air line, $x_O = 0.2095(1 - R_{LU} \cdot x_F)$ for LFC line, and $x_O = \lambda_1 = 0.116$ for MOC line, we have the safe zones displayed an explosive triangle diagram shown in Fig. 6.22.
b. With initial condition $x_F = 0.01875$, $x_O = 0.14665$, we can plot the sample point in explosive triangle diagram (Fig. 6.23).

For a dilution operation, we have three possibilities controlled by different dilution curves.
For a dilution by diluent/air, the dilution curve is $x_F = x_{F,0}$.
For a dilution by fuel/air, the dilution curve is $x_O = 0.2095 \cdot \left(1 - x_F - x_{D,0}\right)$.
For a dilution by diluent/fuel, the dilution curve is $x_O = 0.2095 \left(1 - x_{F,0} - x_{D,0}\right) = 0.14665$.
All three curves are plotted on the flammable envelope. From these operations, we can make the following conclusions.

(a) For a dilution by diluent/air, adding nitrogen will reach ($x_O = 0.116$, $x_F = 0.01875$) for the non-flammable state, and ($x_O = 0.1102$, $x_F = 0.01875$) for the non-ignitable state. They can be derived analytically by solving

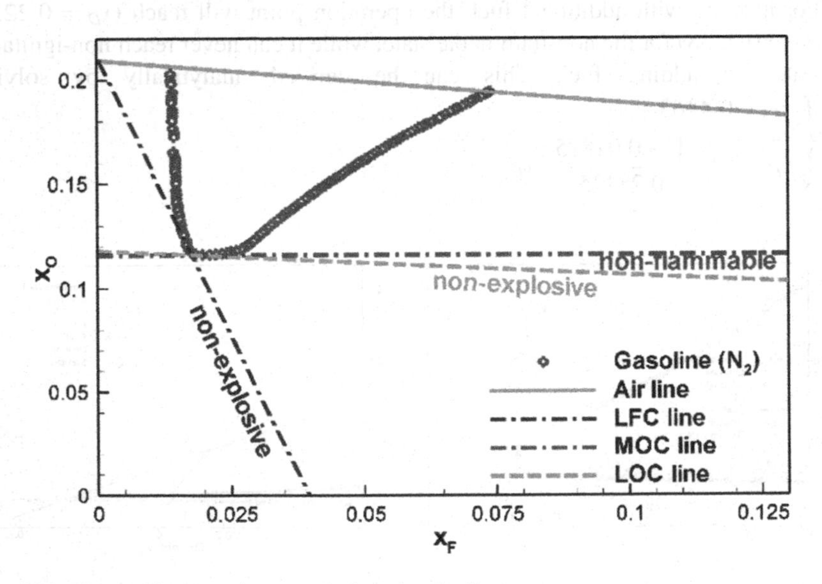

Fig. 6.22 Flammable status in an explosive triangle diagram

$$\begin{cases} x_F = 0.01875 \\ x_O = 0.2095 \cdot (1 - 0.4463) = 0.116 \end{cases} \text{ and } \begin{cases} x_F = 0.01875 \\ x_O = 0.2095 \cdot (1 - 25.27 \cdot x_F) \end{cases}$$

respectively.

(b) For a dilution by fuel/air, adding fuel will reach ($x_O = 0.116$, $x_F = 0.165$) for the non-flammable state, while adding air will reach ($x_D = 0.14815$, $x_F = 0.01159$) for non-ignitable state. They can be derived analytically by solving $\begin{cases} x_O = 0.2095 \cdot (1 - x_F - 0.28125) \\ x_O = 0.2095 \cdot (1 - 0.4463) = 0.116 \end{cases}$ and $\begin{cases} x_D = 0.2095 \cdot (1 - x_F - 0.28125) \\ x_O = 0.2095 \cdot (1 - R_{LU} \cdot x_F - x_F) \end{cases}$ respectively.

(c) For a dilution by diluent/fuel, adding diluent will reach ($x_D = 0.14665$, $x_F = 0.01187$) for the non-flammable state. It can also be determined by solving $\begin{cases} x_O = 0.14665 \\ x_O = 0.2095 \cdot (1 - 25.27x_F) \end{cases}$.

c. The dilution operations can be realized in an explosive triangle diagram shown in Fig. 6.23.

d. With initial condition $x_F = 0.01875$, $x_O = 0.14665$, we can plot the sample point in an explosive triangle diagram (Fig. 6.24).

For a purge operation, we have 3 possibilities controlled by different dilution curves.

For a purge operation by diluent, the purge curve is $x_O = 0.2095 \cdot \left(1 - x_F - x_{D,0}\right) \frac{x_F}{x_{F,0}}$.

For a purge operation by fuel, the purge curve is $x_O = 0.2095(1 - x_{F,0} - x_{D,0}) \cdot \frac{1 - x_F}{1 - x_{F,0}}$.

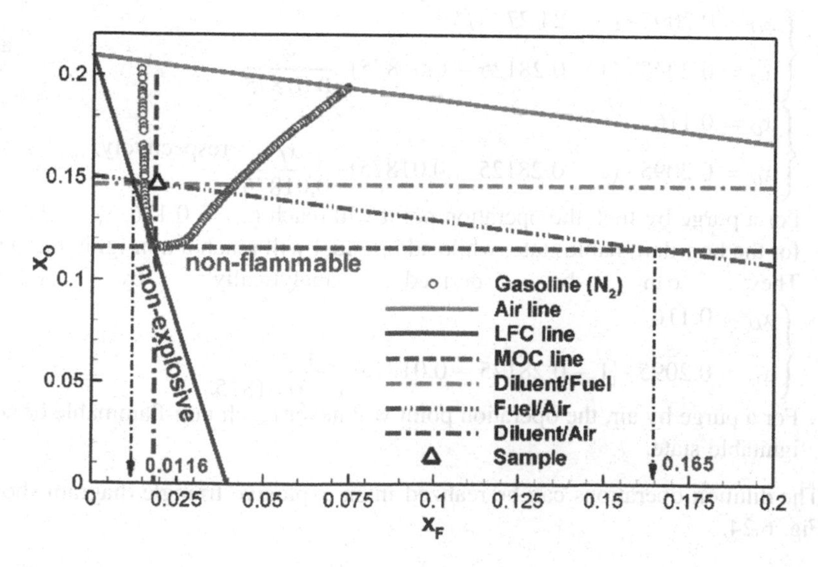

Fig. 6.23 Dilution operations in an Explosive Triangle diagram

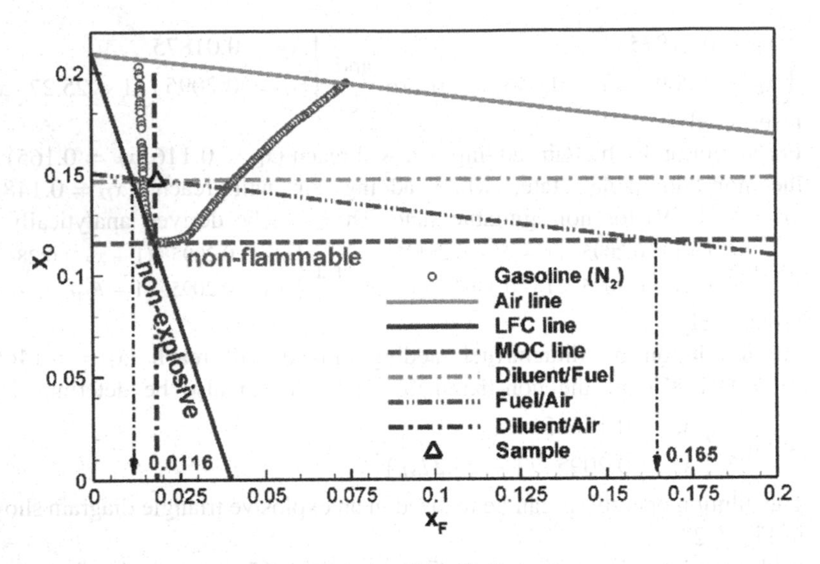

Fig. 6.24 Dilution operations in an Explosive Triangle diagram

For a purge operation by air, the purge curve is $x_O = 0.2095\left[1 - \left(1 + \frac{x_{D,0}}{x_{F,0}}\right) \cdot x_F\right]$.

All three curves are plotted on the flammable envelope. From these operations, we can make the following calculations.

(a) For a purge by diluent, the operation point will reach ($x_O = 0.116$, $x_F = 0.0148$) for the non-flammable state, and ($x_O = 0.1270$, $x_F = 0.0162$) for the non-ignitable state. They can be derived analytically by solving
$$\begin{cases} x_O = 0.2095 \cdot (1 - 24.27 \cdot x_F) \\ x_O = 0.2095 \cdot (1 - 0.28125 - 0.01875) \cdot \dfrac{x_F}{0.01875} \end{cases}$$
and
$$\begin{cases} x_O = 0.116 \\ x_O = 0.2095 \cdot (1 - 0.28125 - 0.01875) \cdot \dfrac{x_F}{0.01875} \end{cases}$$
respectively.

(b) For a purge by fuel, the operation point will reach ($x_O = 0.116$, $x_F = 0.2238$) for the non-flammable state, while adding fuel will reach a non-ignitable state. They can be derived analytically by solving
$$\begin{cases} x_O = 0.116 \\ x_O = 0.2095 \cdot [1 - 0.28125 - 0.01875] \dfrac{1 - x_F}{1 - 0.01875} \end{cases}.$$

(c) For a purge by air, the operation point will never reach non-flammable or non-ignitable state.

The dilution operations can be realized in an explosive triangle diagram shown in Fig. 6.24.

Table 6.5 Target safe operation points via various operations

Operation	Standard	Diluted	Triangle
Dilution (D/A) to non-flammable	$x_F = 0.01875$ $x_D = 0.455$	$x_m = 0.446$ $R = 22.5$	$x_F = 0.01875$ $x_O = 0.116$
Dilution (D/A) to non-ignitable	$x_F = 0.01875$ $x_D = 0.4275$	$x_m = 0.48$ $R = 24.27$	$x_F = 0.01875$ $x_O = 0.01102$
Dilution (F/A) to non-flammable	$x_F = 0.165$ $x_D = 0.28125$	$x_m = 0.446$ $R = 1.8$	$x_F = 0.165$ $x_O = 0.116$
Dilution (F/A) to non-ignitable	$x_F = 0.01159$ $x_D = 0.28125$	$x_m = 0.29$ $\dot{R} = 24.27$	$x_F = 0.01159$ $x_O = 0.14815$
Dilution (F/D) to non-ignitable	$x_F = 0.01187$ $x_D = 0.28813$	$x_m = 0.3$ $R = 24.27$	$x_F = 0.01187$ $x_O = 0.14665$
Purge (D) to non-flammable	$x_F = 0.00918$ $x_D = 0.4583$	$x_m = 0.446$ $R = 29.1$	$x_F = 0.0148$ $x_O = 0.116$
Purge (D) to non-ignitable	$x_F = 0.2238$ $x_D = 0.2225$	$x_m = 0.404$ $R = 24.27$	$x_F = 0.0162$ $x_O = 0.127$
Purge (F) to non-flammable	$x_F = 0.01187$ $x_D = 0.28813$	$x_m = 0.446$ $R = 0.9941$	$x_F = 0.2238$ $x_O = 0.116$

Here is a summary of the above graphical operations in Table 6.5. They are consistent with each other, showing the equivalence of flammability diagrams.

6.7.3 Filling and Emptying a Fuel Tank

Problem 6.3 Here is one problem on gasoline tank safety. If there is a tank of 10 m^3, how much nitrogen is needed to inert the tank before filling the tank? If there is 2 m^3 of liquid gasoline is already stored inside, how much nitrogen is needed to dilute the headspace in the tank before emptying the tank?

Solution:
From the diluted flammability diagram in Fig. 6.25, we can find the critical parameters ($R_{LU} = 24.27$, $x_{LU} = 0.446$).

For a safe filling operation, we need to dilute the oxygen level from 0.2095 to ISOC (or LOC).

$$ISOC = LOC = \lambda_2 = \frac{0.2095(1 - x_{LU})}{1 - \frac{x_{LU}}{1+R_{LU}}} = \frac{0.2095 \times (1 - 0.446)}{1 - \frac{0.446}{1+24.27}} = 0.118$$

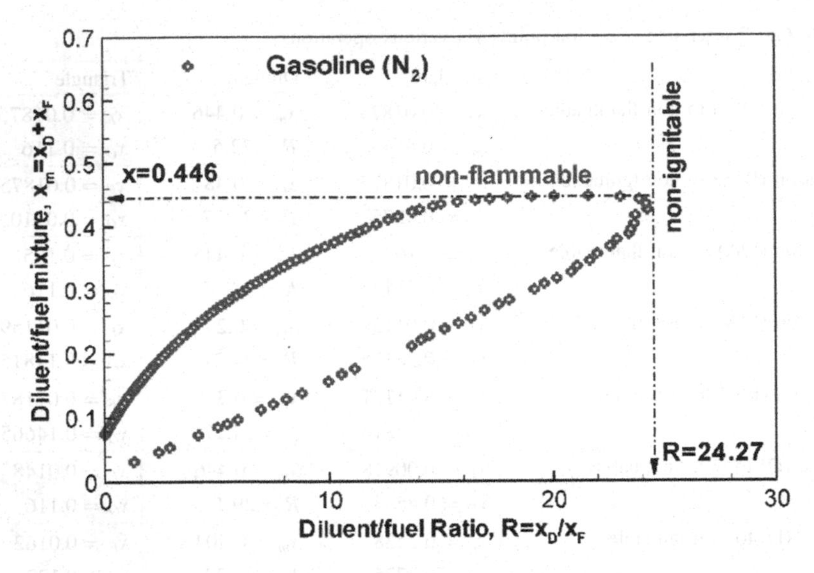

Fig. 6.25 Flammable state of gasoline in a diluted flammability diagram

$V_{N_2} = -V_0 \cdot \ln\left(\frac{x_O}{x_{O,0}}\right) = -10 \times \ln\left(\frac{0.118}{0.2095}\right) = 5.74\ \text{m}^3$. That means 5.74 m^3 of nitrogen is needed to purge the empty space in the tank.

For a safe emptying operation, we need to dilute the fuel concentration from 100 % to OSFC (or LFC), which is estimated as:

$$OSFC = LFC = \beta_2 = \frac{1}{1 + R_{LU}} = \frac{1}{1 + 24.27} = 0.0396$$

$V_{N_2} = -V_0 \cdot \ln\left(\frac{x_F}{x_{F,0}}\right) = -(10 - 2) \times \ln\left(\frac{0.0396}{1}\right) = 25.83\ \text{m}^3$. That means 25.83 m^3 of nitrogen is needed to dilute the headspace in the tank.

Chapter 7
Applications on Fuel Streams (Type II Problem)

Recent news on a gas explosion in Taiwan (August 1, 2014) shows the ignition potential of accidental discharge of flammable gas is still not fully recognized. This chapter covers type II flammability. Type II flammability problems cover intertizing fuel streams to a point where no danger of ignition is possible if the stream is released to atmosphere. Specifically, only two types of problems are solved in this chapter, one, how to burn diluted fuel safely for clean combustion technologies and two, how to inert a fuel stream against potential ignition of leaking flammable refrigerant. In both cases mentioned, the fuel streams are released into air, where the mixture explosibility is fixed and cannot be modified. Therefore, we have to control the ignitability of the stream instead. This method deals with difusion flames, so it is fundamentally different from type I problem.

7.1 Clean Combustion Technologies

As environmental pressures mount on industrial companies to further decrease the amount of hydrocarbons released to the atmosphere the combustion of low calorific value gases (LCVG) are becoming increasingly important. The combustion of low calorific value gases (LCVG) is becoming increasingly important as the environmental pressure is mounting on releasing them directly into the atmosphere. Used in many industrial processes but without clear definition, 'low calorific value gases' (LCVG) or 'low-BTU Gas' is usually associated with the product of air-blown coal gasifier which typically have higher heating values of less than 7 MJ/m^3 (188 BTU/ft^3) [117]. Some industrial processes, such as coal gasification to make LCVG, are the one of the cheapest ways to eliminate sulfur and substantially reduce nitric oxide emissions both from fuel bound and air nitrogen. Fuel dilution can cause many problems, such as a reduced flame temperature, and consequently, burning rate, narrow stability limits and low combustion efficiencies. Since, dilution causes a narrowing of the flammable zone we must make up for this by other means. To increase the flammability zone after dilution we can either use high temperature preheating or adopt a high calorific value gas. Adding a high calorific value gas into the fuel stream will also improve flame stabilization. The flammability problem of LCVG is a diluent-fuel mixture burning in normal air.

T. Ma, *Ignitability and Explosibility of Gases and Vapors*,
DOI 10.1007/978-1-4939-2665-7_7

Though controlled combustion for industrial purpose has been adopted for centuries, only recently has a motivation for research into clean combustion technologies increased. The research of combustion regimes is greatly expanded for better efficiencies and lower emissions. To utilize widely available diluted fuels (or LCVG), generally there are three types of solutions, oxy-combustion, high temperature air combustion (HTAC), and hydrogen-doping. In order to reduce O_2 emission from coal-fired power generation, oxy-fuel combustion is proposed to use pure oxygen or a mixture of O_2 and recycled flue gas for generating high CO_2 concentration product gas. Oxy-fuel capture shows advantages over post-combustion capture in terms of capture cost of CO_2. For oxy-fuel combustion, the concern is the limiting oxygen concentration that will not support flame propagation. Since the oxygen isolation process is commonly expensive, another solution is to burn the diluted fuel with another fuel, commonly hydrogen. However, there is no theory on why using hydrogen instead of ethane or propane, which is much cheaper. A third solution is preheating or high-temperature air combustion [118], which is actively researched as a low-NOx combustion technology recently.

The safe operation on LCVG depends on combustion stability, or specifically the flammability of fuels. As pointed out by Chomiak et al. [117], although the flammability limits cannot be directly applied for evaluating whether or not a LCVG can be burned in a given device, generally practical systems operate at temperature in excess of the adiabatic flame temperature at the lower flammable limits. Therefore, the flammability limits may serve as a first approximate of the condition necessary for burning a mixture. Though Muniz and Mungal [119] suggested the local flow velocity must be near the premixed laminar burning velocity as one criterion for stabilization, another criterion is the composition of the fuel/oxidizer mixture must be within the flammability limits. So flammability limits are still needed for combustion stability reasons.

Historically, the flammability of mixed gases has been solved by Jones and Kennedy [120] adopting the famous Le Chatelier's rule. Recently numerical simulations tools, such as GRI3.0 and CHEMKIN, or premixed flame code from Sandia National Lab [121] are used. They produced some useful information about the burning of diluted methane, however, these methods are too complex for estimation purposes. Most empirical rules are still based on Le Chatelier's rule, which treat the role of diluent as part of a pseudo fuel. But pairing fuel/diluent together into pseudo fuels, the flammability limits can be estimated (Coward [68], Burgess et al. [122], Heffington [123]). Beyond these methods, systematic research on flammability is still missing, probably due to the difficulty in experiments and the limitation from a flammability theory.

7.1.1 Role of Diluent on Flammability

For CO_2-diluted methane as a typical LCVG, the flammability problem is expressed as diluted fuel burning in normal air. Here the added diluents will change the thermal signature of the fuel. For the diluted flammability diagram, the energy conservation equation at LFL is $x_F \cdot Q_F + x_D \cdot Q_D + 1 - x_L = x_F H_F$, so we have the thermal balance at critical flammability limits as shown below.

$$\frac{x_L}{1+R} \cdot Q_m + \frac{R \cdot x_L}{1+R} \cdot Q_D + (1 - x_L) \cdot 1 = \frac{x_L}{1+R} \cdot C_O H_O \tag{7.1}$$

$$\frac{x_U}{1+R} Q_m + \frac{R \cdot x_U}{1+R} Q_D + (1 - x_U) \cdot 1 = \lambda \cdot (1 - x_U) \cdot H_O \tag{7.2}$$

where $Q_m = Q_F \cdot \beta + Q_D \cdot (1 - \beta)$ is the quenching potential of the diluted fuel, $H_m = H_O$ is unchanged, since the fuel type is the same. The total energy release is scaled down in $C_m = C_O \cdot \beta$.

Solve the above equations, we have the flammability envelope bounded by

$$x_L = \frac{1}{1 + \frac{C_m H_m}{1+R} - \frac{Q_m}{1+R} - \frac{Q_D R}{1+R}} \tag{7.3}$$

$$x_U = \frac{\lambda \cdot H_m - 1}{\lambda \cdot H_m - 1 + \frac{Q_m}{1+R} + \frac{Q_D \cdot R}{1+R}} \tag{7.4}$$

Forcing $x_L = x_U$, we have a crossing point (R_{LU}, x_{LU}) which is the theoretical inertion point.

$$R_{LU} = \frac{C_m H_m \cdot \lambda - \lambda Q_m - H_m}{\lambda Q_D} \tag{7.5}$$

where additive rule applies for $C_m = C_O \cdot x_F$, $H_m = H_O \cdot x_F$, $Q_m = Q_F \cdot \beta + Q_D \cdot (1 - \beta)$. Forcing $R_{LU} = 0$ (or equivalent to say, shrinking the envelope to nil), we have the critical fuel concentration as

$$\beta = \frac{\lambda Q_D}{\lambda C_O H_O - C_O - \lambda \cdot (Q_F - Q_D)} \tag{7.6}$$

Submit parameters for methane/CO_2 mixture, $Q_D = 1.75$, $\lambda = 0.2095$, $H_O = 16.4$, $Q_F = 13.8$, we have the critical fuel concentration as $\beta = 0.1562$. That is the limiting methane concentration with CO_2 dilution. If the concentration drops to this value, the flammability envelope shrinks to nil and the mixture is non-explosive.

A series of flammable envelopes are displayed in Fig. 7.1a. With an increasing dilution level, both flammability limits in air are raised. However, the flammability ratio is smaller, meaning the flammability zone is narrower, as shown in Fig. 7.1b. An interesting observation is that the inertion point stays the same level as the fresh fuel. That means, the dilution of the fuel will not change the minimal oxygen concentration (MOC), which is a property of background air, or explosibility.

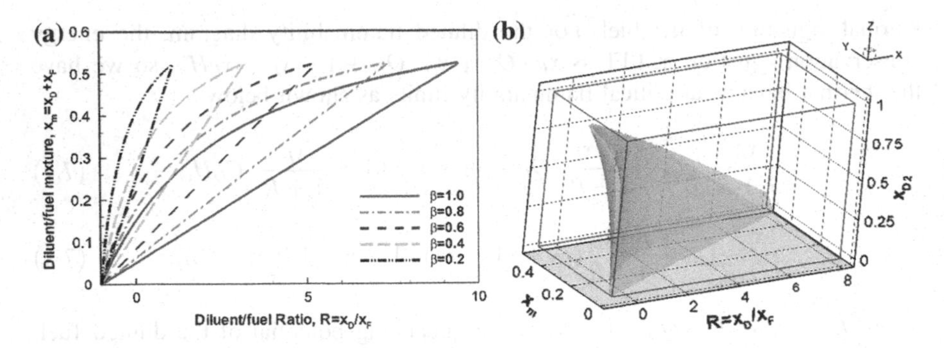

Fig. 7.1 Role of CO_2-dilution on flammable envelope in **a** diluted flammability diagram **b** 3-D HQR diagram

7.1.2 Role of Oxygen on Flammability (Oxy-combustion)

For oxy-combustion, we have another form to present the role of oxygen on the flammable envelope, which is controlled by two curves developed in Sect. 5.1.1.

$$x_L = \frac{Q_m}{Q_m + \frac{C_O H_O}{1+R} - \frac{Q_F}{1+R} - \frac{Q_D R}{1+R}} \tag{7.7}$$

$$x_U = \frac{\lambda \cdot H_O - Q_m}{\lambda \cdot H_O - Q_m + \frac{Q_F}{1+R} + \frac{Q_D R}{1+R}} \tag{7.8}$$

The flammable envelopes modified by oxygen level are presented in figure (a), while its HQR diagram is shown in Fig. 7.2b.

Chen et al. [121] simulated the flammable region of methane in O_2/N_2 and O_2/CO_2 using the well-stirred reactor (WSR) model and the detailed reaction mechanism GRI-Mech 3.0 with an adiabatic thermal boundary condition in CHEMKIN.

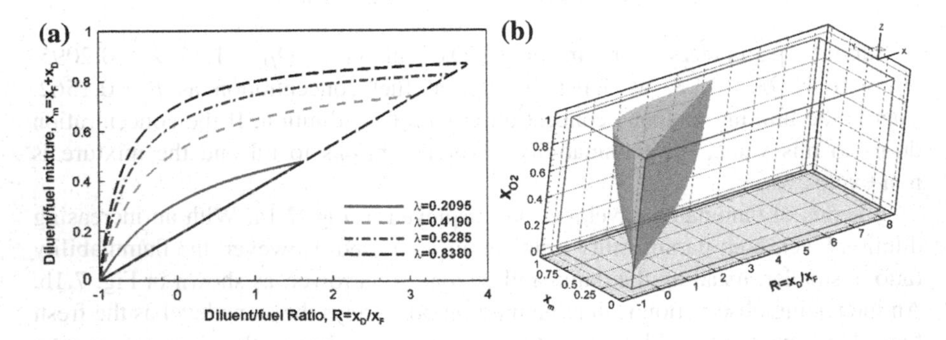

Fig. 7.2 Role of oxygen on flammable envelope in **a** diluted flammability diagram **b** 3-D HQR diagram

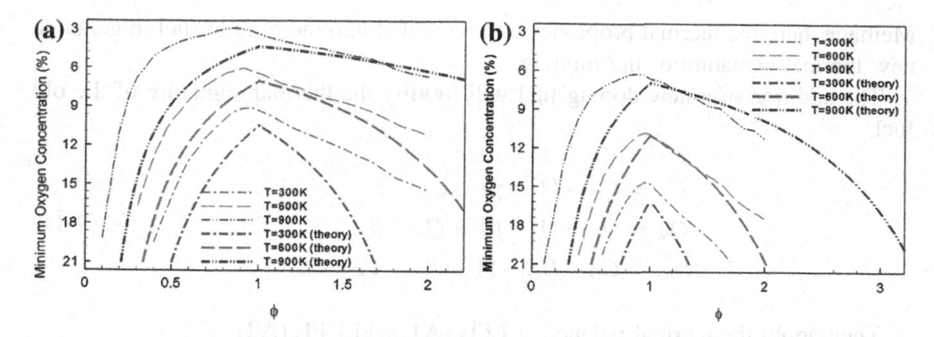

Fig. 7.3 Flammability region prediction from WSR and thermal balance method. **a** Nitrogen. **b** Carbon dioxide

The dashed lines in Fig. 7.3 show the flammability regions under different ambient temperature conditions with a residence time of 0.1 s. Blowout occurs when the operating conditions is outside the flammability region.

Here the flammability problem for oxy-combustion is a high-temperature fuel burning in an oxygen-modified background air. The thermal balance method can produce a similar region in thick-solid lines, which are generated by finding the flammability limits (Eqs. 7.7 and 7.8) at a certain oxygen level.

$$x_L = \frac{\lambda \cdot Q_m}{\lambda \cdot Q_m + C_O H_O - \eta Q_F} \tag{7.9}$$

$$x_U = \frac{\lambda \cdot H_O - \eta Q_m}{\lambda \cdot H_O - \eta Q_m + \eta Q_F} \tag{7.10}$$

Figure 7.3 shows the flammability zones for the fuel (Methane) at various ambient temperatures. In order to facilitate understanding, the fuel concentration is converted to the equivalence ratio. The simulation result of Chen et al. [121] are also displayed as a comparison. Without experimental data, it is difficult to judge which method is better. However, this method is a hand-calculation method, while the WSR model is a numerical tool. This method is much cheaper in terms of computational cost and demand on external data support.

7.1.3 Role of Hydrogen Doping on Flammability (Hydrogen-Doping Combustion)

With the narrowed flammability region caused by the dilution, a common choice of expanding the flammability zone is adding a high-calorific value gas to the mixture. For diluted methane as a fuel, hydrogen is commonly chosen for its high heat value, strong diffusivity, as well as other parameters. Similar to the role of CO_2 to

Methane, here the thermal properties will be added into the pseudo fuel to generate new thermal signature of the mixture.

The addition of a new doping fuel will modify the thermal signature of the old fuel.

$$
\begin{aligned}
C_m &= C_{O1} \cdot (1-\gamma) + C_{O2} \cdot \gamma \\
Q_m &= Q_{F1} \cdot (1-\beta) + Q_{F2} \cdot \beta \\
H_{Om} &= C_{O1} \cdot H_{O1} \cdot (1-\beta) + C_{O2} \cdot H_{O2} \cdot \gamma
\end{aligned}
\tag{7.11}
$$

Then apply the thermal balance at LFL (A1) and UFL (A2)

$$\frac{x_L}{1+R} \cdot Q_m + \frac{R \cdot x_L}{1+R} \cdot Q_D + (1-x_L) \cdot 1 = \frac{x_L}{1+R} \cdot C_O H_O \tag{7.12}$$

$$\frac{x_U}{1+R} Q_m + \frac{R \cdot x_U}{1+R} Q_D + (1-x_U) \cdot 1 = x_{O2} \cdot (1-x_U) \cdot H_O \tag{7.13}$$

So the new flammable envelope is bounded by the following two curves.

$$x_L = \frac{1}{1 + \frac{C_m H_m}{1+R} - \frac{Q_m}{1+R} - \frac{Q_D R}{1+R}} \tag{7.14}$$

$$x_U = \frac{\lambda \cdot H_m - 1}{\lambda \cdot H_m - 1 + \frac{Q_m}{1+R} + \frac{Q_D \cdot R}{1+R}} \tag{7.15}$$

By varying the hydrogen concentration, we can have the modified flammability diagram shown in Fig. 7.4. Here $\beta = 0.3$ means it is a fuel mixture of 30 % Methane and 70 % CO_2. $\gamma = 0.06$ means a 6 % hydrogen-doping.

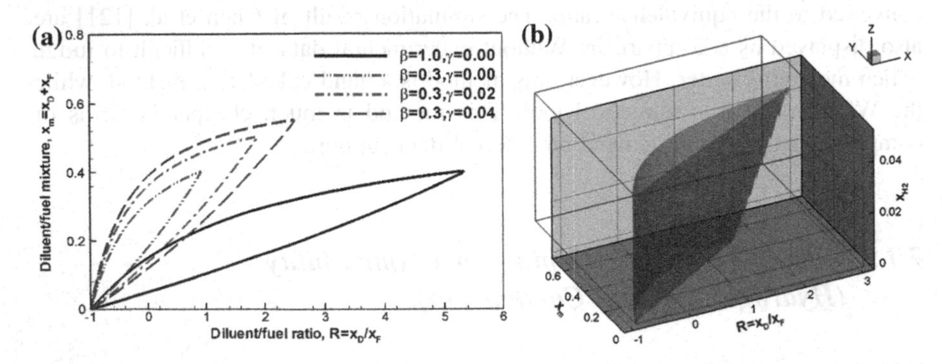

Fig. 7.4 Role of hydrogen-doping on flammable envelope for Methane/CO_2 mixtures in **a** diluted flammability diagram **b** 3-D HQR diagram

7.1.4 Role of Temperature on Flammability (High Temperature Combustion)

The temperature-dependence of thermal properties is established upon the enthalpy relationships between species provided by NIST chemistry webbook. Reconstructing a simpler correlation for air, we have the enthalpy of air.

$$E_{air} = \left(H^0_{AFT} - H^0_{298.15}\right)_{air} = f(T) = 1.4893\mathrm{T}^2 + 29.862\mathrm{T} - 9.381 \qquad (7.16)$$

Note the input is T/1000, so the coefficients can have more valid digits. Next, define a temperature-dependent enthalpy-scaling factor $\eta(T)$ for measuring the system enthalpy change in reference to air.

$$\eta(T) = \frac{E_i}{E_{air}} = \frac{\left(E^{1600}_{air} - E^{T}_{air}\right)}{\left(E^{1600}_{air} - E^{298}_{air}\right)} = \frac{42.21 - \left(1.4893\mathrm{T}^2 + 29.862\mathrm{T} - 9.381\right)}{42.21 - (-0.35)}$$
$$= 1.212 - 0.035 \cdot T^2 - 0.702 \cdot T \qquad (7.17)$$

For the thermal balance at critical limits, we have

$$\left(\frac{x_L}{1+R} \cdot Q_F + \frac{R \cdot x_L}{1+R} \cdot Q_D + 1 - x_L\right) \cdot \eta = \frac{x_L}{1+R} \cdot C_O H_O \qquad (7.18)$$

Rearrange the terms, we have a new equation for LFL with R as the only input.

$$x_L = \frac{\eta}{\eta + \left(\frac{C_O H_O}{1+R} - \frac{Q_F \cdot \eta}{1+R} - \frac{Q_D \cdot R \cdot \eta}{1+R}\right)} \qquad (7.19)$$

Similarly, we can solve

$$\left(\frac{x_U}{1+R} Q_F + \frac{R x_U}{1+R} Q_D + 1 - x_U\right) \cdot \eta = 0.2095(1 - x_U) H_O \qquad (7.20)$$

Sometimes, the stable burning of the diluted fuel is accomplished through preheating. With a temperature factor η, the flammability equations are updated as

$$x_L = \frac{\eta}{\eta + \left(\frac{C_O H_O}{1+R} - \frac{Q_F \cdot \eta}{1+R} - \frac{Q_D \cdot R \cdot \eta}{1+R}\right)} \qquad (7.21)$$

$$x_U = \frac{\lambda \cdot H_O - \eta}{\lambda \cdot H_O - \eta + \left(\frac{Q_F}{1+R} + \frac{Q_D \cdot R}{1+R}\right) \cdot \eta} \qquad (7.22)$$

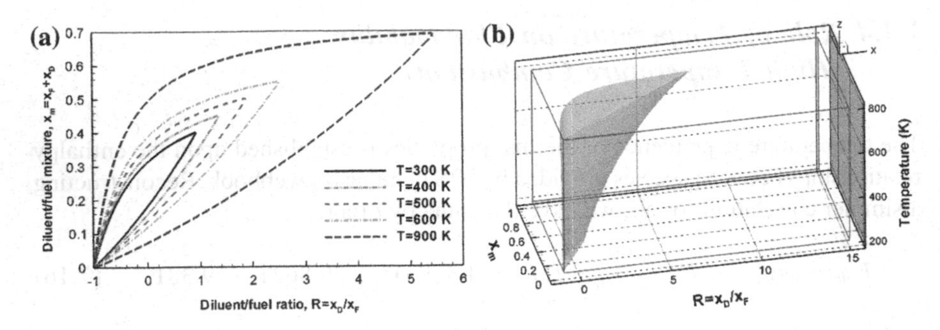

Fig. 7.5 Temperature dependence of the flammable envelopes in **a** diluted flammability diagram **b** 3-D HQR diagram

Forcing $x_L = x_U$, we have the cross point

$$R_{LU} = \frac{C_O H_O - \eta Q_F - C_O \eta}{\eta Q_D} \tag{7.23}$$

Using the thermal properties of the fuel/diluent/oxygen, we can set up the temperature-modified flammability diagram shown in Fig. 7.5.

7.1.5 Discussion on Fuel Properties

Currently in the U.S. market Ethane is as cheap as methane in terms of calorific value, so why not use Ethane for doping? See Fig. 7.6 for examples of simulations conducted with Ethane and Propane.

Figure 7.6 shows that 10 % of ethane or propane will expand the explosibility (increasing the inerting R, or LOC is increased by doping since net fuel is more), but not the flammability (x_m stays roughly constant, or MOC is unchanged by

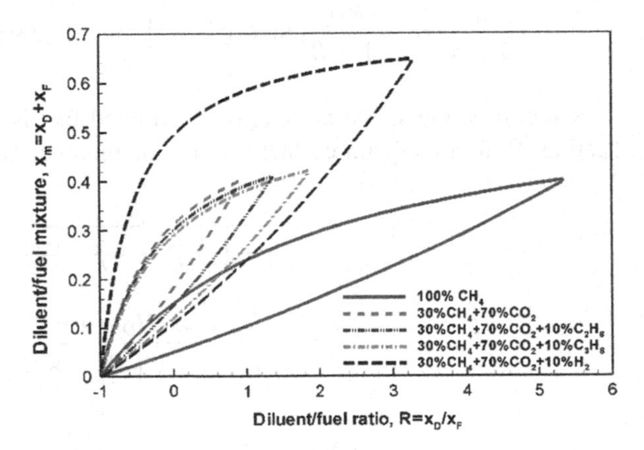

Fig. 7.6 Role of doping with different fuels

doping since the fuel properties are similar). Adding Ethane/Propane will enhance the explosibility (net fuel LHV, lower heating value, is increased), but will not affect the energy releasing capability of oxygen which is fuel specific, so UFL is almost the same.

Doping with ethane or propane will affect LFL (specifically lower LFL, or expand explosibility) only, since they belong to a same family of fuels with methane. However, they can be used as the supporting fuel for High-Temperature Air Combustion (HTAC), which will change the flammable envelope by raising the air temperature. Numerous research work has been performed and published without knowing this theoretical flammability diagram, for example, the monograph on high-temperature combustion [118].

What is the best candidate fuel for doping? Literature gives no clue. It is commonly recommended that a fuel with a larger heat of oxidation is a better choice. Heat of oxidation is the amount of energy release per unit mole of oxygen. Any fuel with a better oxygen-based energy release will serve as the doping fuel for expanding the flammability zone.

From the process for constructing the flammability diagram, the ratio of heat releasing potential (H_O) over heat absorbing potential (Q_F) is a good index for the purpose of expanding the flammability region. A larger ratio means this fuel contributes more energy and absorbs little energy by mass. Table 7.1 lists some fuels with a larger ratio, which may serve as candidates for the doping purpose on improving the flammability zone in burning diluted fuels.

Comparing with methane's ratio of 1.19, hydrogen has a ratio of 15.68, while the best fuel (pentaborane) has a ratio of 290.94. Indeed, pentaborane can expand the flammability zone much larger than hydrogen, since it is highly reactive and used as a rocket fuel. A comparison of some potential fuels are provided in Fig. 7.7.

Table 7.1 Candidate fuels for enhancing the combustion stability

Fuels	C_O	x_L	x_U	Q_F	H_O	H_O/Q_F
Pentaborane	9.74	0.42	98	0.08	24.35	290.915
Diborane (g)	4.50	0.84	98	0.09	26.27	285.825
Urea (s)	1.50	0.9	98.0	0.29	73.65	250.099
Methylhydrazine (monomethylhydrazine)	2.50	2.5	97	0.07	15.64	222.141
Dichlorosilane	1.50	4.7	96	0.08	13.58	176.667
Unsymmetrical dimethylhydrazine (UDMH)	4.00	2.3	95	0.06	10.64	164.45
Chloroprene	5.00	4.0	20	0.03	4.81	160.551
Di-*tert*-butyl peroxide (DTBP)	11.49	0.8	88	0.17	10.80	62.7076
Trichlorosilane	0.50	7.0	83	1.02	28.64	27.9662
Formaldehyde (g)	1.00	7.0	73	0.72	14.01	19.5745
Carbon monoxide (g)	0.50	12.5	74	0.80	15.60	19.5726
Arsine (g)	0.25	5.1	78	5.39	96.07	17.8082
Deuterium (g)	0.50	4.9	75	2.77	44.38	16.0458
Hydrogen	0.5	4.0	75.0	3.51	55.02	15.6802

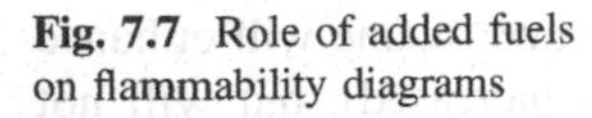

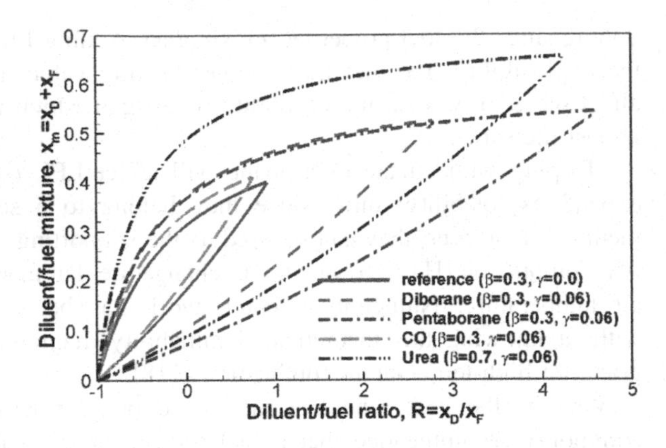

Fig. 7.7 Role of added fuels on flammability diagrams

Figure 7.7 shows the performance of some candidate fuels on enlarging the flammable envelope. Surprisingly, hydrogen is not the best performer while urea has a better performance on expanding the envelope. The impact of CO-doping is trivial, which is unique to CO only and perhaps related to the special CO chemistry. Generally, the contribution of a fuel is not only dependent on its own thermal signature, but also on the thermal properties of the primary fuel and ambient air. Some fuels may have a better performance than hydrogen. To be utilized at industrial scale, they must be subject to other selecting criteria, such as availability and cost. Flammability diagram provides a useful tool for checking their performance against a certain mixing combination.

7.2 Critical Flammability Ratio

The Bureau of Mines showed an early interest on the flammability of refrigerants when American mines were developed deep enough requiring the protection of refrigerants [124]. Attention to the flammability of refrigerants arises again as a result of the ban of Halons and the phase-out of hydrochloroflorocarbons (HCFCs). Even in this post-Halon era, an ideal refrigerant fluid to meet all property requirements cannot be found. Instead, mixtures of fluids are proposed to generate the best properties for the replacement of HCFCs [110, 111]. Different HFCs offer favorable characteristics, such as thermodynamic properties, performance, and compatibility with existing equipment, moderate toxicity and relatively low cost comparable with HCFCs. However, mixing of HFCs may induce the problem of accidental explosion if not treated properly. So the flammability of mixtures has both theoretical and practical value to the design of an ideal refrigerant blend.

In order to understand the flammability of a mixture, standard flammability diagram [125], diluted flammability diagram [105] and the ternary diagram [110, 126] were developed to demonstrate the change of flammability envelope upon mixing. However, a consistent theory to interpret this change is still missing.

Thus the understanding of mixture flammability is fragmental and not complete. After 100 years, Le Chatelier's rule is still the core theory behind the mixture flammability [105].

For inerting a dynamic fuel stream, the concept of critical flammability ratio (CFR) is introduced to guide the mixing of refrigerants. Generally, the critical flammability ratio of a blend is the critical ratio of the sum of the flammable component molar (volumetric) concentrations to the sum of the inert (nonflammable) component concentrations. So CFR $= 1/R_{LU}$. Thus, the CFR yields the maximum allowed concentration of flammable component necessary to obtain an overall nonflammable refrigerant blend formulation in air at a defined temperature and pressure.

Figure 7.8 shows how to read R_{LU} from the diluted flammability diagram. By definition, CFR is the critical fuel/diluent ratio, or the reverse of R_{LU}. This diagram can be reorganized into standard flammability diagram, as shown in Fig. 7.9. Note the dilution lines in Fig. 7.9 are controlled by $R_{LU} = \frac{x_D}{x_F}$.

Finally, we can reorganize the data into the explosive triangles shown in Fig. 7.10. The dilution line is controlled by $x_O = x_{LOC2} = 0.2095[1 - (1 + R_{LU}) \cdot x_F]$, while the inertion line is controlled by MIC lines.

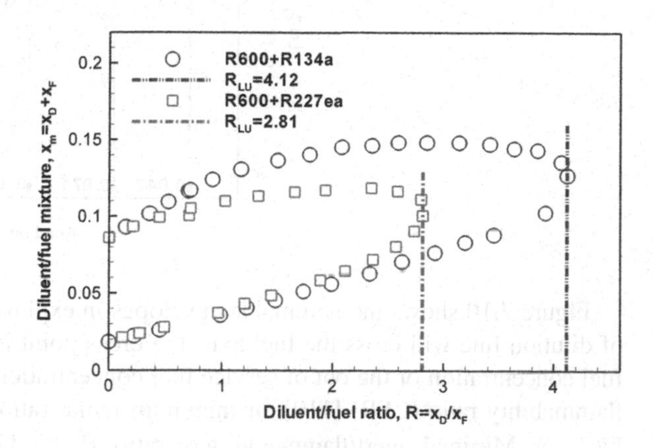

Fig. 7.8 Critical flammability ratio in diluted flammability diagrams

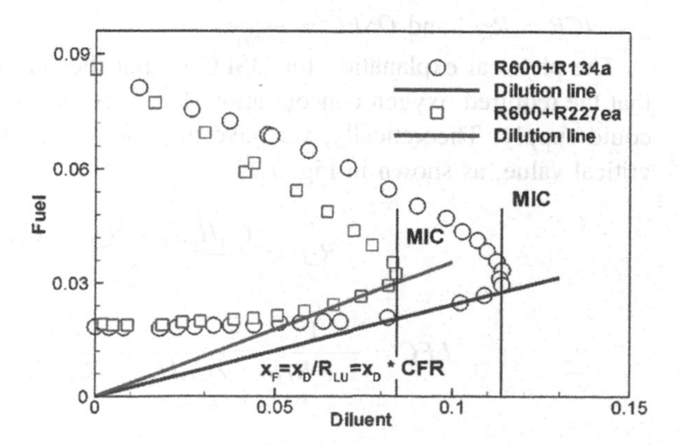

Fig. 7.9 Critical flammability ratio in standard flammability diagrams

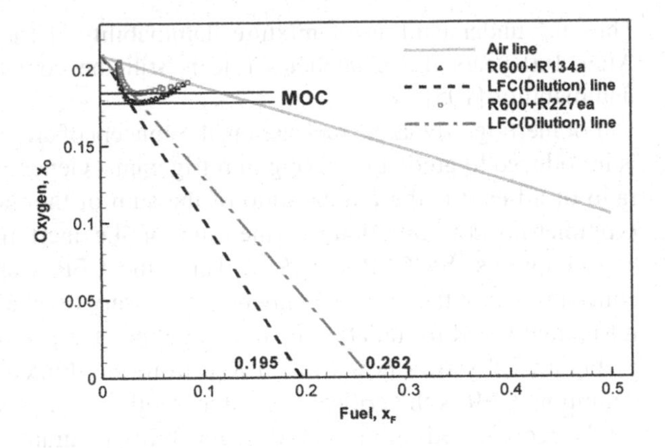

Fig. 7.10 Critical flammability ratio in explosive triangle diagrams

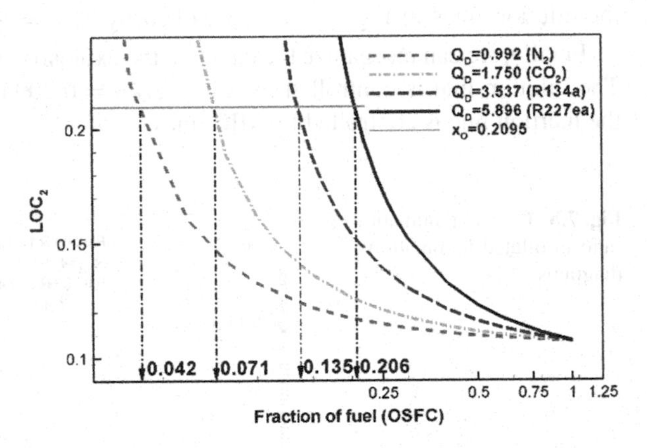

Fig. 7.11 LOC curves modified by various diluents

Figure 7.10 shows the flammable envelopes in explosive triangle. The extension of dilution line will cross the fuel axis, the cross point is called the critical diluted fuel concentration or the out of service fuel concentration(OSFC) [125], the critical flammability ratio (CFR) [110], or minimum molar ratio (fuel to diluent)(1/MMR) [70], or Minimal inert/flammable gas ratio (ICR) [70]. Note, $CFR = \frac{1}{MMR} = \frac{1}{R_{LU}}$, $ICR = R_{LU}$, and $OSFC = \frac{1}{1+R_{LU}}$.

The physical explanation for OSFC is that the fuel concentration is so diluted that the required oxygen concentration (LOC) is raised above what the ambient air could supply. Theoretically, we have the following equations for predicting this critical value, as shown in Fig. 7.11.

$$R_{LU} = \frac{C_m H_m \lambda - \lambda Q_m - C_m}{\lambda Q_D} \tag{7.24}$$

$$LFC = \frac{1}{1 + R_{LU}} = \frac{\lambda Q_D}{\lambda C_O H_O - C_O - \lambda(Q_F - Q_D)} \tag{7.25}$$

Figure 7.11 shows LOC curves modified by various diluents. When the background air cannot supply enough oxygen to fully consume the diluted fuel, the fuel stream is fully diluted. Sometimes, it is also called inerted [125]. Here dilution is preferred instead of inertion. Inertion is realized mainly by mass, while dilution is realized mainly by volume. Here the diluent (R227ea) is mainly working on volume (molar) basis. On mass basis, the diluent requirements are significantly different among various diluents. This fact is in sharp contrast to an inertion problem, where the diluent mass is roughly constant between various diluents [14]. In other words, inertion is controlled by mass (of the diluent), while dilution is controlled more by volume. They are different concepts, which should be clarified within a flammability diagram.

7.3 Problems and Solutions

Here are examples on type II problems, inerting a fuel stream.

Problem 7.1 If a stream of diluted methane (30 % of fuel with 70 % of nitrogen, 1 m^3/s) is released into the air. How much nitrogen is further needed to inert the stream? [79]
Solution:

1. Using Table 4.2, find the MMR/MMF of methane, MMR $= 6.005$, MMF $=$ 0.4231
2. Find LFC, LFC $= 1/(1+\text{MMR}) = 1/(1+6.005) = 0.143$
3. Here the fuel concentration is 30 %, which needed to be diluted to 0.143.
4. Solving $LFC = 14.3\,\% = \frac{0.3\cdot\dot{V}_F}{\dot{V}_F+\dot{V}_D} = \frac{1\times0.3}{1+\dot{V}_D}$, we have $\dot{V}_D = 1.10\,\text{m}^3/\text{s}$
5. In addition, MFC $= 1/(1+\text{MMR}) = 1/(1+6.005) = 0.0605$
6. That means, when the fuel stream is released in air, its fuel concentration will drop from 0.143 to 0.0605, still non-ignitable. The above calculations are displayed in Fig. 7.12.

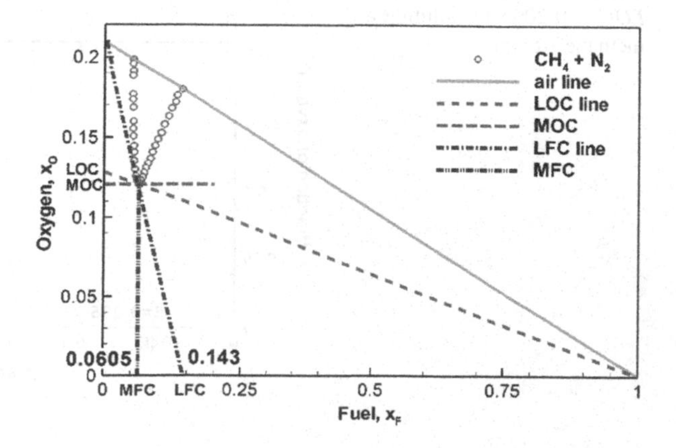

Fig. 7.12 Flammability diagrams for diluting a methane stream

Problem 7.2 If there is a methane leak from a 100 % methane pipeline at a rate of 2 L/s. How much diluent (L/s) is needed to dilute the methane stream so the mixture is non-ignitable? (using theoretical flammability diagram).

Solution:

1. Find the thermal properties of methane: $C_O = 2, x_L = 0.05, x_U = 0.15$
2. The thermal signature of methane is found to be $H_O = 16.4$ and $Q_F = 13.8$.
3. For dilution by nitrogen ($Q_D = 0.992$), we have

$$LFC = \frac{\lambda Q_D}{\lambda C_O H_O - C_O - \lambda \cdot (Q_F - Q_D)}$$
$$= \frac{0.2095 \times 0.992}{0.2095 \times 2 \times 16.4 - 2 - 0.2095 \times (13.8 - 0.992)} = 0.095$$

In order to dilute the methane stream from 100 to 9.5 %, the nitrogen stream is found by solving $OSFC = 9.5\,\% = \frac{\dot{V}_F}{\dot{V}_F + \dot{V}_D} = \frac{2}{2 + \dot{V}_D}$, or $\dot{V}_D = 19.05$ L/s.

4. Similarly, for dilution by CO_2 ($Q_D = 1.75$), we have

$$LFC = \frac{\lambda Q_D}{\lambda C_O H_O - C_O - \lambda \cdot (Q_F - Q_D)}$$
$$= \frac{0.2095 \times 1.75}{0.2095 \times 2 \times 16.4 - 2 - 0.2095 \times (13.8 - 0.992)} = 0.156$$

In order to dilute the methane stream from 100 to 9.5 %, the nitrogen stream is found by solving $OSFC = 15.6\,\% = \frac{\dot{V}_F}{\dot{V}_F + \dot{V}_D} = \frac{2}{2 + \dot{V}_D}$, or $\dot{V}_D = 10.82$ L/s.

A graphical explanation of the result is shown in Fig. 7.13.

Problem 7.3 Here is an example about CO safety. There is a CO leakage with 10 L/s, how much diluent (N_2) is needed to inert this stream? Without an experimental flammability diagram, redo your estimations based on theoretical flammability diagrams.

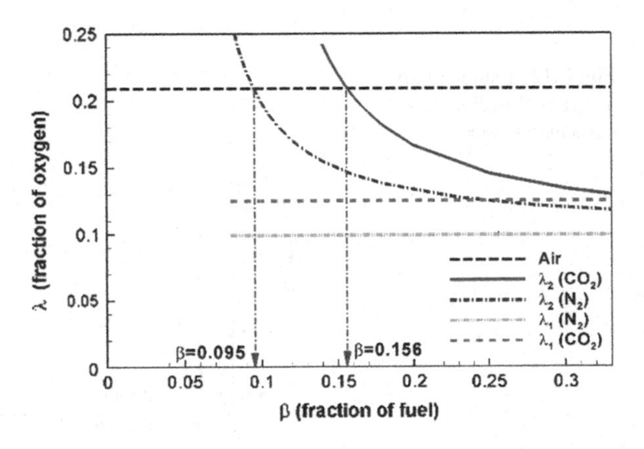

Fig. 7.13 LFC as a result of $LOC = 0.2095$ for diluting a methane stream

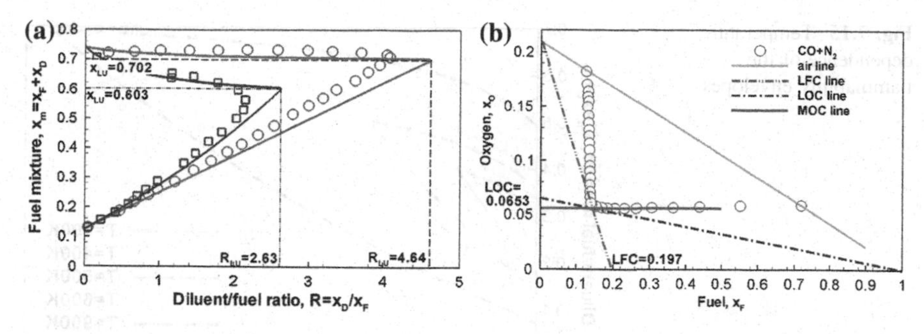

Fig. 7.14 Flammability diagrams for CO safety. **a** Diluted flammability diagram. **b** Explosive triangle diagram

Solution:

This problem can be solved in two ways, using experimental data and using theory.

(a) Using experimental data, the out-of-service fuel concentration (LFC or OSFC) is 0.197 in the explosive triangle diagram (as shown in Fig. 7.14b). Therefore, the diluting requirement for inerting the fuel stream is $\dot{V}_{N_2} = \frac{\dot{V}_{fuel}}{OSFC} - \dot{V}_{fuel} = \frac{10}{0.197} - 10 = 40.76\,\text{L/s}$.

(b) Without experimental data, the theoretical inertion point for nitrogen (shown in Fig. 7.14a) is determined as

$$R_{LU} = \frac{C_O H_O \lambda - \lambda Q_F - C_O Q_m}{\lambda Q_D} = \frac{0.2095 \times 0.5 \times 15.59 - 0.2095 \times 0.8 - 0.5 \times 1}{0.2095 \times 0.992} = 4.64$$

$$x_{LU} = \frac{Q_m}{Q_m + \frac{C_O H_O}{1+R_{LU}} - \frac{Q_F}{1+R_{LU}} - \frac{Q_D R_{LU}}{1+R_{LU}}} = \frac{1}{1 + \frac{0.5 \times 15.59}{1+4.64} - \frac{0.8}{1+4.64} - \frac{0.992 \times 4.64}{1+4.64}} = 0.702$$

Then the critical concentration is determined as

$$LFC = \frac{1}{1+R_{LU}} = \frac{1}{1+4.64} = 0.177$$

So the diluting requirement for inerting the fuel stream is $\dot{V}_{N_2} = \frac{\dot{V}_{fuel}}{OSFC} - \dot{V}_{fuel} = \frac{10}{0.177} - 10 = 46.5\,\text{L/s}$.

Problem 7.4 Here is an example on temperature dependency. A certain process is releasing methane diluted by carbon dioxide. The fuel fraction in this stream is 0.3. In order to ignite and burn this diluted fuel @ 16 %, what is the minimal temperature for preheating the fuel stream to be ignitable?

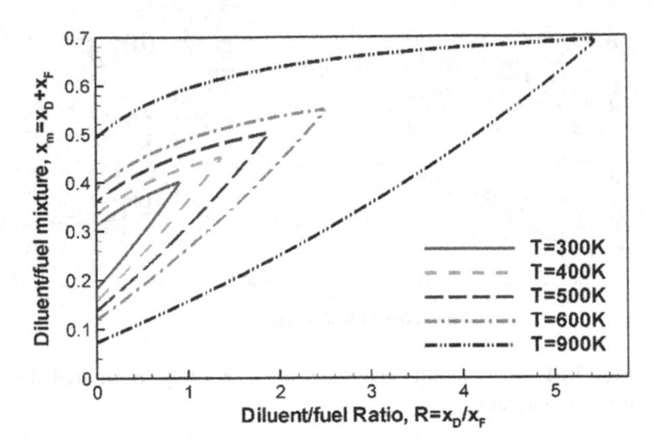

Fig. 7.15 Temperature dependence of the flammability envelopes

Solution:

Step 1. Find the thermal signature of the fuel using the spreadsheet below.

	x_i	C_O	x_L	x_U	Q_F	H_O
CO_2	0.70	–	–	–	1.75	–
CH_4	0.30	2.00	5.00	15.00	13.81	16.40
Mixture	–	0.60	–	–	5.37	16.40

Step 2. Using the thermal properties of the diluted fuel we can set up the temperature-modified flammability diagram shown in Fig. 7.15.

Step 3. From this diagram, we can find a raised temperature of 400 K is enough to lower the lower flammable limit to 0.16.

Another solution: Solve the conversion factor η directly from the

$$x_L = \frac{\eta}{\eta + \left(\frac{C_O H_O}{1+R} - \frac{Q_F \cdot \eta}{1+R} - \frac{Q_D \cdot R \cdot \eta}{1+R}\right)} = 0.16 = \frac{1}{1 + \frac{C_O H_O}{\eta} - Q_F} = \frac{1}{1 + \frac{0.6 \times 16.4}{\eta} - 5.367}$$

$$\eta = 0.92655$$

$$\eta(T) = \frac{E_i}{E_{air}} = \frac{\left(E_{air}^{1600} - E_{air}^{T}\right)}{\left(E_{air}^{1600} - E_{air}^{298}\right)} = \frac{42.21 - \left(1.4893\mathrm{T}^2 + 29.862\mathrm{T} - 90.381\right)}{42.21 - (-0.35)}$$

$$= 1.212 - 0.035 \cdot T^2 - 0.702 \cdot T = 0.92655$$

$$T^2 + 20.06T - 8.156 = 0$$

$$\mathrm{T} = 0.415$$

So the answer is 415 K.

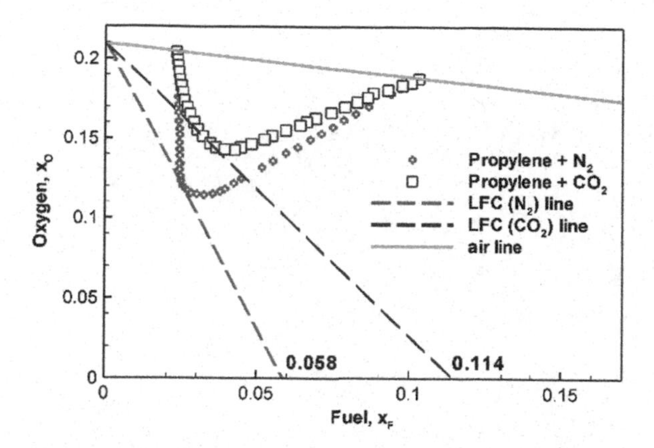

Fig. 7.16 LFC in the explosive triangle diagram for propylene

Problem 7.5 If there is a propylene gas leak similar to the gas explosion in Taiwan, ROC, and the leak rate is 1 kg/min, how much nitrogen is needed for dilution? How much carbon dioxide is needed? What is the difference in mass flow rate?

Solution:

From the measured diagrams in Fig. 4.8, we can have an explosive triangle diagram for propylene reconstructed below. From the tabled critical parameters in Table 4.2, we have $R_{LU} = 15.14$ for nitrogen inertion and $R_{LU} = 7.735$ for carbon dioxide inertion. This translate to LFC $= 0.058$ and LFC $= 0.114$ respectively (as shown in Fig. 7.16).

The molecular weight of propylene is 44 g/mol.
The volumetric flow rate is $1000(g/min)/44(g/mol)^{*}22.4(L/mol) = 509.1$ L/min.
The dilution requirement for nitrogen is

$$\dot{V}_{N_2} = \frac{\dot{V}_{fuel}}{LFC} - \dot{V}_{fuel} = \frac{509.1}{0.058} - 509.1 = 8268.5 \text{ L/min.}$$

The dilution mass requirement for nitrogen is

$$\dot{m}_{N_2} = \frac{8268.5}{22.4} \times 28 = 10335 \text{ g/min} = 10.3 \text{ kg/min}$$

The dilution requirement for carbon dioxide is

$$\dot{V}_{CO_2} = \frac{\dot{V}_{fuel}}{LFC} - \dot{V}_{fuel} = \frac{509.1}{0.114} - 509.1 = 3956.7 \text{ L/min.}$$

The dilution mass requirement for carbon dioxide is

$$\dot{m}_{CO_2} = \frac{3956.7}{22.4} \times 44 = 7772 \text{ g/min} = 7.77 \text{ kg/min}$$

Chapter 8
Applications in Compartment Fires (Type I Problem)

A compartment fire is characterized by its variable oxygen level inside the compartment. This leads to the non-explosive zone overlapping with the non-ignitable zone. The standard procedure for treating a compartment fire is to set aside the oxygen as part of normal air, then treat the leftover as a flammable air-free mixture burning in air. A typical compartment fire problem is an underground coal mine fire. Since Sir Humphrey Davy's time, the explosion potential of flammable gases in a confined space is a constant research topic. This flammable mixture is characterized by multiple fuels, multiple diluents, reduced oxygen level and above-ambient temperature. During the combustion process, various flammable gases are generated in flaming or smoldering processes, complicating the situation of a variable oxygen level. Fuels are typically dealt with either Le Chatelier's rule. Without a consistent theory to treat variable oxygen, explosive triangles are resorted to find the flammability of fuel vapor in air, rather than a modified oxygen/nitrogen background (self-flammability). Various improvements were proposed to better present flammability of a mixture. Cheng et al. [85] summarize 6 graphical methods to determine the mine gas explosibility, ternary diagram, revised Le Chatelier's Rule, Maximum allowable oxygen (MAO) analysis, USBM method, Coward's explosive triangle and Kukuczka method. Ternary diagram is the standard diagram proposed by Zabetakis [43], Crowl [86], Schroeder et al. [127]. Kukuczka method [128] is a variation of Coward explosive triangle, which is also the standard method for analyzing mine gases [129−131]. USBM explosibility diagram is a popular method used in mine industry [43], with a targeted methane-air-inert range. Revised Le Chatelier's Rule has been adopted in estimating inerting concentrations [132]. In practice, explosive triangle (for oxygen content) plus Le Chatelier's Rule has been widely used to determine the mixture explosibility since 1948, and still being improved for better presentation effects [133].

Comparing with mine fires, the critical fire behaviors in compartment fires, such as backdraft and blowtorch effect, are attracting more attentions recently. Though the exact triggering event may be different, the nature of these critical behaviors is explosive burning induced by mixing ignitable but non-explosive fuel mixture with non-ignitable but explosive air. Experience from studies on mine fires can be tailored to understand these critical behaviors. Lacking a consistent theory to treat the role of oxygen, most flammability diagrams are treating the flammable gases as

T. Ma, *Ignitability and Explosibility of Gases and Vapors*,
DOI 10.1007/978-1-4939-2665-7_8

a fuel burning in normal air. Various schemes are proposed to improve the representativeness of some indexes, but not from the first principles. Here with the flexibility of the thermal balance method, and the intuitiveness of flammability diagrams, we can have a better look at type I problems, including mine fires and compartment fires.

8.1 Underground Mine Fires

Underground coal fire problems are becoming a serious environmental disaster [134] due to their duration and uncontrollable state. Uncontrolled coal mine fires have contributed a significant share of carbon dioxide to the atmosphere, enhancing the potential of global warming [135]. China, India, and US are major coal consumers, suffering more from the burden of underground coal fires [136]. Generally, the control of underground coal mine fires is realized through sealing, i.e., stopping the supply of oxygen. However, due to the complexity in underground structures, the oxygen supply cannot be easily controlled, which calls for a gas diagnosis for explosibility control. Numerous efforts are put into the detection, diagnosis, and suppression of such fires [135]. One of the challenges in diagnosing the underground fire is that the fire is sealed underground with limited information available. Safety regulation requires that the air samples directly from active underground areas should be taken and analyzed on a regular basis. As long as the problem area continues to be ventilated, certain gases, gas ratios, and equations can be used to determine the status of burning. Morris [137] discusses the technical aspects of gas sampling and analysis methodologies. In order to extract the information from the gas sampling results, several indices are proposed to get some information from the sampled species information. According to Timko [131], these indices are proposed to answer four questions

1. If a fire exists
2. What is burning
3. What is the status of fire after sealing?
4. If the sealed atmosphere is explosive when ventilation commences?

For the first question, the presence of carbon monoxide or carbon dioxide [138] is a good indicator of a fire. For the second question, Jones-Trickett Ratio [139] determines the sample reliability. For the third question, the power industry engineers used the relative intensity equations [140] to characterize the quantity of air available to burn a unit mass of fuel, the percentage of oxygen consumed, and the effect of temperature on temperature. Litton [141] developed Litton's ratio to monitor a sealed mine atmosphere. In another paper [142], Litton proposed the R-index as X_{CO}/X_D^2, which is a ratio of fuel/diluent, checking the ignitability (fuel) of the mixture. Realizing that low molecular weight hydrocarbon concentrations increase with rising temperature, Justin and Kim [143] developed the hydrocarbon ratio to describe the status in a mine fire. For the last problem, Zabetakis et al. [144] modified Coward's diagram, which examined explosibility by comparing methane and oxygen. They also proposed the Maximum Allowable Oxygen (MAO) analysis to determine if the atmosphere is inert or explosive. Without a reasonable understanding of the thermal nature of each constituent, they convert all fuels into methane, all diluent into nitrogen, to draw the graph. Cheng et al. [145] refined Coward's explosion triangle to generate an explosibility safety factor for assessing the explosion risk. Ray et al. [146] applied the above indices to assess the status of sealed fire in underground coal mines. Singh et al. [147] gave some case studies on applying these indices.

More than one century of experimental and theoretical work on flammability stresses the ternary (fuel/oxygen/diluent) picture of a flammability problem, making a universal graphical method difficult to apply and explain. How to reliably predict the role of fuel/diluent/oxygen on the flammable envelope requires a detail analysis of the thermal balance within a mixture. So a thermal balance method [15] adopting a binary view on each species has been proposed. All species has a mass, with a non-zero quenching potential, while fuel or oxygen can release energy, with LFL controlled by fuel and UFL controlled by oxygen. This binary treatment makes the contribution of each species cumulative and additive, thus making a hand-calculation possible. This method has been applied to find the flammability of a flammable mixture. Now this method will be tailored to reconstruct the flammable envelope under various oxygen levels, with a special emphasis on analyzing the flammable gas mixtures from a coal-mine fire.

8.1.1 Oxygen-Modified Flammability Diagram

On May 13, 1940, a fire was discovered in the Continental mine of the Hazel Brook Coal Co. Centralia, PA [129]. This is a well-instrumented, well-documented and thoroughly analyzed fire, where typical samples of the mine fire atmosphere are listed in Table 8.1. Each mixture sample is composed of two diluents, three fuels, plus oxygen.

Table 8.1 Recordings of the mine-fire atmosphere during an underground coal fire

Date, 1940	19-May	21-May	24-May	24-May	4-Jun	5-Jun	5-Jun	5-Jun	5-Jun	7-Jun	8-Jun
Sample	1	2	3	4	5	6	7	8	9	10	11
CO_2	9.53	3.70	6.50	6.90	12.30	13.00	12.60	12.50	11.10	2.20	14.00
O_2	8.13	15.30	11.50	10.20	2.90	0.40	0.20	0.70	4.30	18.00	2.60
CO	0.91	0.60	1.10	1.40	1.00	4.50	6.60	6.60	2.30	0.30	0.40
CH_4	0.08	0.00	0.00	0.00	0.10	0.90	1.00	0.80	0.00	0.10	1.00
H_2	0.41	0.60	1.50	1.90	1.50	7.00	8.80	9.90	2.70	1.00	8.30
N_2	80.94	79.80	79.40	79.50	82.20	74.20	70.80	69.50	79.60	78.40	73.70

Table 8.2 Input data for deriving thermal signature of different species

	C_o	X_L	X_U	Q_F	H_o
CO_2	–	–	–	1.75	–
O_2	–	–	–	1.06	–
CO	0.5	12.5	74.0	0.80	15.59
CH_4	2.0	5.0	15.0	13.81	16.40
H_2	0.5	4.0	75.0	3.51	55.02
N_2	–	–	–	0.99	–

As shown in Sect. 7.1.2, Oxygen level is affecting the upper flammable boundary. In order to reconstruct a flammability diagram, the thermal signature of each component should be determined first. Following the procedure in thermal balance method in Sect. 4.1, the thermal signatures of all components are either retrieved (for a diluent) or derived (for a fuel). Fuels get their thermal signature from LFL/UFL information, while the quenching potentials of diluents and oxygen are directly computable from their enthalpy information (retrievable and constant in reference to air enthalpy, i.e., the original definition, see Eq. 3.2). Then the thermal signature of this mixture is estimated by summing-up energy terms. Here the thermal signature of all species is provided and listed in Table 8.2.

Flammability limits are estimated by reversely converting the energy terms into the concentration domain. Flammable envelopes outlined by LFL and UFL lines can also be reconstructed from the thermal signature of this mixture.

Here is a typical example for reconstructing diluted flammability diagrams. For the sample taken on May 19, 1940, it has the following constituents: 9.53 % CO_2, 8.13 % O_2, 0.91 % CO, 0.08 % CH_4, 0.41 % H_2, and 80.94 % N_2. You are required to

a. reconstruct the flammability diagram to determine the flammable state of this mixture.
b. redraw the Coward's explosion triangle for the flammable state of this mixture.

Solution:

Part A. This is a flammability problem within a variable-oxygen environment. The air is composed of 8.13 % O_2 + 80.94 % N_2, the fuel is composed of 0.08 % CH_4, 0.41 % H_2, 0.91 % CO. The diluent is 9.53 % CO_2. In order to reconstruct the flammability diagram, the following steps are necessary.

1. The oxygen fraction in the air is 8.13/(8.13 + 80.94) = 9.13 %. Therefore, the quenching potential of air is $Q_m = \lambda \cdot Q_O + (1 - \lambda) \cdot Q_N = 0.997$.
2. The quenching potential of the fuel mixture is $C_O = 0.586$, $Q_D = 2.33$, $H_O = 25.61$

Fuel	Fraction	x_i (%)	C_O	x_L	x_U	Q_F	H_O
CO	0.91	65.00	0.5	12.5	74	0.80	15.59
CH_4	0.08	5.71	2	5	15	13.81	16.40
H_2	0.41	29.29	0.5	4	75	3.51	55.02
Sum	1.40	100	0.586			2.33	25.61

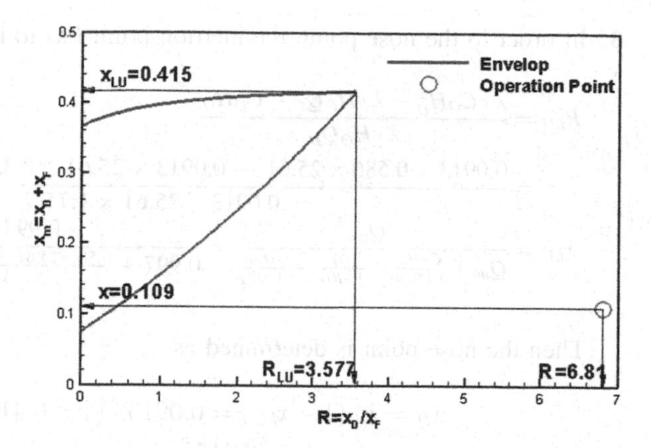

Fig. 8.1 Flammable states of the mine-fire atmosphere before the active fire

3. The operation point is $x_m = x_F + x_D = 0.014 + 0.0953 = 0.1093, R = 9.53/1.4 = 6.81$.
4. Using $Q_D = 1.75$ for carbon dioxide, reconstruct the flammability diagram using the above inputs in Eqs. 7.14 and 7.15.
5. Since the operation point is outside the flammability envelope (see Fig. 8.1), this mixture is not flammable (too lean to burn).

Part B. This is a flammability problem in a variable-oxygen atmosphere. MOC is used for the nose value in the Coward's explosion triangle.

1. Find the LFL point as

$$x_L(R=0) = \frac{Q_m}{Q_m + \frac{C_O H_O}{1+R} - \frac{Q_F}{1+R} - \frac{Q_D R}{1+R}} = \frac{0.997}{0.997 + 0.586 \times 25.61 - 2.33}$$
$$= 0.0729$$

$$x_O = \lambda \cdot (1 - x_L) = 0.0913 \cdot (1 - 0.0729) = 0.0846$$

2. Find the UFL point as

$$x_U = \frac{\lambda \cdot H_O - Q_m}{\lambda \cdot H_O - Q_m + \frac{Q_F}{1+R} + \frac{Q_D \cdot R}{1+R}} = \frac{0.0913 \cdot 25.61 - 0.993}{0.0913 \cdot 25.61 - 0.993 + 2.33} = 0.366$$

$$x_O = \lambda \cdot (1 - x_U) = 0.0913 \cdot (1 - 0.366) = 0.0579$$

3. In order to the nose point, the inertion point has to be determined first.

$$R_{LU} = \frac{\lambda \cdot C_O H_O^2 - \lambda \cdot H_O Q_F - C_O H_O}{\lambda \cdot H_O Q_D}$$
$$= \frac{0.0913 \times 0.586 \times 25.61^2 - 0.0913 \times 25.61 \times 2.33 - 0.586 \times 25.61}{0.0913 \times 25.61 \times 1.75} = 3.5766$$
$$x_{LU} = \frac{Q_m}{Q_m + \frac{C_O H_O}{1+R_{LU}} - \frac{Q_F}{1+R_{LU}} - \frac{Q_D R}{1+R_{LU}}} = \frac{0.997}{0.997 + \frac{0.586 \times 25.61 - 2.33 - 1.75 \times 3.5766}{4.5766}} = 0.4155$$

Then the nose point is determined as

$$x_O = \lambda \cdot (1 - x_{LU}) = 0.0913 \cdot (1 - 0.4155) = 0.0533$$
$$x_F = \frac{x_{LU}}{1+R} = \frac{0.4155}{1 + 3.5766} = 0.091$$

4. The operation point from measurement data is $x_F = 0.014$, $x_O = 0.0813$
5. With the above four points, we can reconstruct the Coward's explosive triangle diagram as below (Fig. 8.2).

For the gases from a mine fire, all fuels (CO, H_2, CH_4) are grouped together as a pseudo fuel, while nitrogen and oxygen are grouped together as a pseudo air. Then the carbon dioxide is treated as the diluent. From the thermal properties of the pseudo fuel/air/diluent, we can determine the flammable state by comparing oxygen levels.

$$\begin{cases} \lambda < \lambda_1 \rightarrow \text{non-flammable} \\ \lambda < \lambda_2 \rightarrow \text{non-explosive} \\ \lambda < \lambda_2 \rightarrow \text{non-ignitable} \end{cases} \tag{8.1}$$

Here for a confined space, non-explosive and non-ignitable are equivalent concepts. Applying the oxygen criteria in Eq. 8.1 to all samples, we can have the status of each sample listed in Table 8.3 and compared in Fig. 8.3.

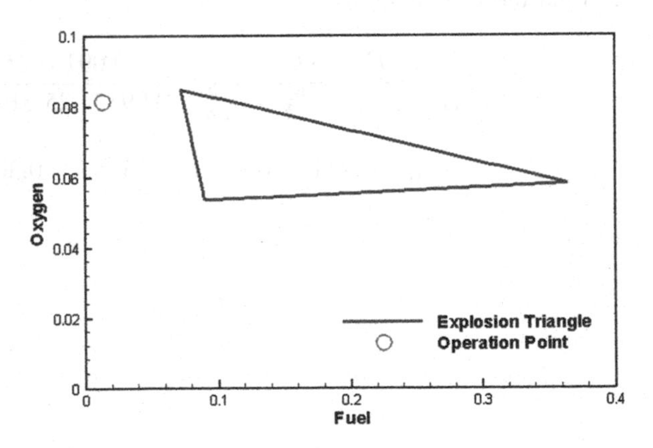

Fig. 8.2 Flammable state of a mine-fire sample before the active fire

Table 8.3 Flammable state of samples from a coalmine fire

	Oxygen	LOC1	LOC2	Flammability	Explosibility	Ignitability
1	0.081	0.10728	0.042972	Non-flammable	Explosive	Ignitable
2	0.153	0.101318	0.03228	Flammable	Explosive	Ignitable
3	0.115	0.091742	0.029714	Flammable	Explosive	Ignitable
4	0.102	0.090822	0.029728	Flammable	Explosive	Ignitable
5	0.029	0.094508	0.02901	Non-flammable	Non-explosive	Non-ignitable
6	0.004	0.015931	0.0305	Non-flammable	Non-explosive	Non-ignitable
7	0.002	0.009645	0.031349	Non-flammable	Non-explosive	Non-ignitable
8	0.007	0.026094	0.030068	Non-flammable	Non-explosive	Non-ignitable
9	0.043	0.087427	0.030741	Non-flammable	Explosive	Ignitable
10	0.180	0.069755	0.026526	Flammable	Explosive	Ignitable
11	0.026	0.047244	0.023329	Non-flammable	Explosive	Ignitable

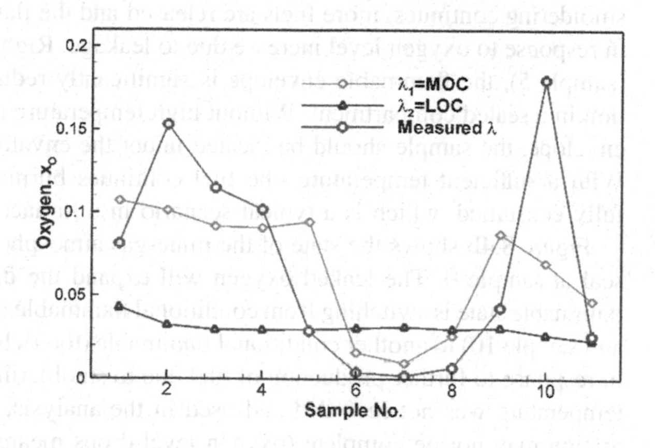

Fig. 8.3 Critical oxygen criteria for determining flammable status of all samples

Figure 8.3 shows the oxygen criteria for determining the flammable state of each sample. If $\lambda < \lambda_2$, the mixture is non-explosive or non-ignitable, which means the mine is actively burning inside. Otherwise, the mine is experiencing smoldering, so there is less fuel produced than oxygen leaked-in. Figure 8.3 also shows the measured oxygen concentrations in reference to theoretical LOCs. MOC is a concept of flammability, as it is derived from and controlled by oxygen and diluent. LOC is a concept of explosibility, so it is a function of the fuel only. If the oxygen level drops below MOC, the reaction is stopped because of insufficient oxygen or too much diluent to sustain the flame propagation. If the oxygen drops below LOC, the reaction is stopped due to insufficient oxygen to match the fuel, and the flammable envelope is non-existent. This is the reason that we cannot make flammability diagrams for samples 6–8.

Following the procedure outlined in Sect. 8.1.1, we can reconstruct the diluted flammability diagrams shown in Fig. 8.4. Note, sample 1–5 were taken before a period of explosive burning at sample 6–8. After that, sample 9–11 were taken when the major burning has been over and the ventilation was resumed and then stopped.

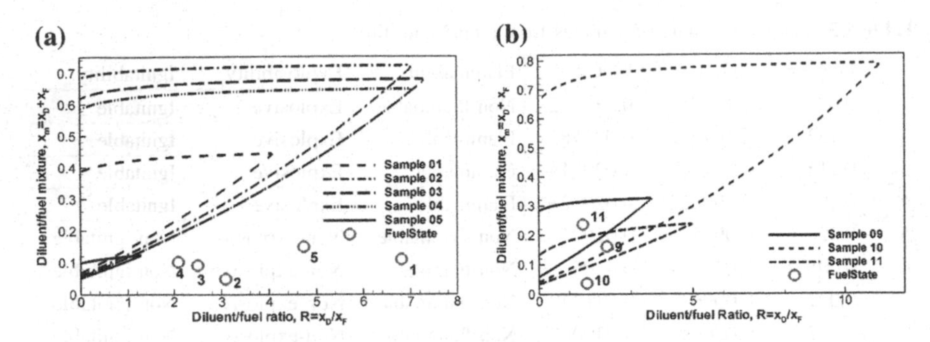

Fig. 8.4 Flammable states of the mine-fire atmosphere. **a** After the active burning. **b** After sealing

Figure 8.4a shows the flammable states before the active burning. Initially, the sample is too lean to burn at sample 1 (non-flammable but explosive). Then as smoldering continues, more fuels are released and the flammable envelope expands in response to oxygen level increase due to leakage. Right before the active burning (sample 5), the flammable envelope is significantly reduced due to oxygen depletion in a sealed compartment. Without high temperature to expand the flammability envelope, the sample should be located under the envelope since the fire is sealed. With a sufficient temperature, the fuel continues burning until fuel or oxygen is fully consumed, which is a typical scenario in a furnace.

Figure 8.4b shows the state of the mine-gas atmosphere after the opening of the seal at sample 9. The leaked oxygen will expand the flammability envelope. The flammable state is switching from conditional flammable (too lean to burn, sample 09 and sample 10) to another conditional flammable (too rich to burn, sample 11) again, in response to further production of fuel due to smoldering. Unfortunately, the local temperature was not recorded and used in the analysis, so the flammability information may not be complete (oxygen level drops means the fire is burning again, so the temperature is rising. However, no temperature information is recorded.).

From the above two diagrams, we can see that the oxygen level determines the flammable envelope size (mostly through lifting the UFL line), while the fuel production is reflected on the position of the operation points. The flammable state of a mixture is depending on the relative position between the envelope (mainly controlled by oxygen) and the sample point (mainly controlled by fuel).

Following procedures in the previous example, we can reconstruct the explosive triangle diagrams shown in Fig. 8.5. They are equivalent to the diluted diagrams in Fig. 8.4.

8.1.2 A Global Progress Variable

One of the challenges in controlling an underground coal mine fire is that it is fully sealed with limited information available through gas sampling. Most of previous

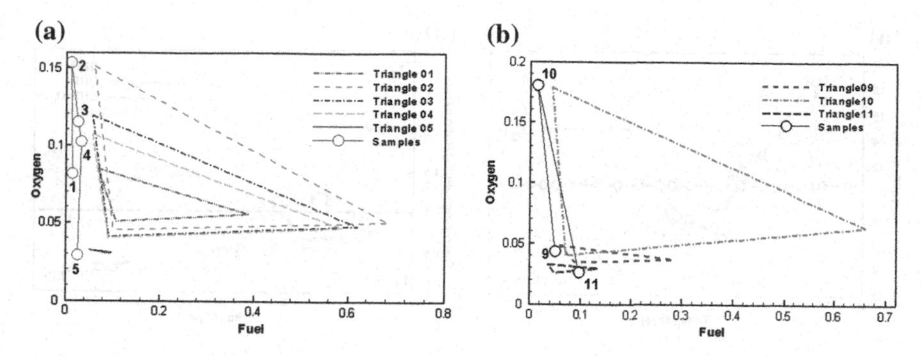

Fig. 8.5 Flammable state representation using Coward explosive triangles. **a** After the active burning. **b** After sealing

works look at the fuel content of the mixture, or the explosibility. A better choice is a progress variable to check both fuel and oxygen for diagnosing what is the status of fire after sealing, and when the sealed atmosphere is explosive if a leaking is allowed.

In essence, the flammability problem of mine gases is a mixture problem of multiple fuels, with carbon dioxide as the diluent, in an oxygen-modified air [148]. The dimensionless variable, Heating/Quenching Ratio, is representative of the competition between heating and quenching.

$$\begin{aligned} HQR_1 &= \frac{C_O H_O x_F}{Q_F x_F + Q_O x_O + Q_D x_D} \\ &= \frac{\text{fuel heating potential}}{\text{mixture quenching potential}} \sim \textit{ignitability} \end{aligned} \quad (8.2)$$

$$\begin{aligned} HQR_2 &= \frac{H_O \cdot x_O}{Q_F x_F + Q_O x_O + Q_D x_D} \\ &= \frac{\text{oxidizer heating potential}}{\text{mixture quenching potential}} \sim \textit{explosibility} \end{aligned} \quad (8.3)$$

$$HQR = \min(HQR_1, HQR_2) \sim \text{flammability} \quad (8.4)$$

If HQR1 > 1, the mixture is ignitable. Note here ignitable has a broad definition, characterized by LFL line, not LFC line. Similarly, explosiveness is characterized by UFL line, not LOC line. That means If HQR2 > 1, the mixture is explosive. If $HQR = \min(HQR_1, HQR_2) > 1$, the mixture is both ignitable and explosive, or flammable, which is most dangerous. The utility of HQR diagram in defining the flammable envelope is shown in Figs. 6.5 and 6.6.

Here they are applied to any mixtures for understanding the flammable/explosive status of a sealed fire. It can also be used to find the flammable envelope in a flammability diagram. It is a global variable reflecting the fundamental energy balance in a gaseous mixture.

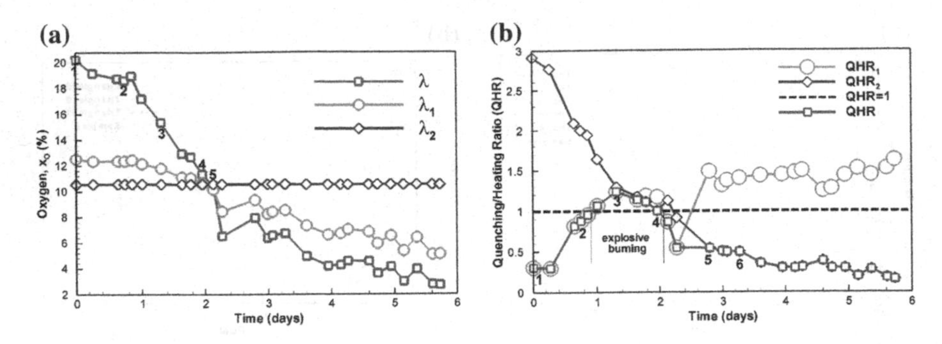

Fig. 8.6 Global explosibility indices for the mixture in the sealed mine. **a** Oxygen criteria. **b** QHR criteria

The first fire to be analyzed happened in China [145]. It was detected in the morning at 6:40 am, sealed though Longwall panel at 8:40 pm, nitrogen injected at 23:00 pm, and the gas sampling began at 10 am the next day. The clock in Fig. 8.6 started then. Compared with normal underground coal mine fires (see next problem), this fire has no hydrogen produced (maybe due to insufficient heat accumulation or low flame temperature), little CO production (around 50 ppm) and C_2H_6 production (less than 33 ppm), so they are omitted here. Note this fire is consistent with the early stage of the next fire, so it is a typical scenario for a sealed fire in its early stages.

The explosibility problem of mine gases is a mixture problem of multiple fuels, with carbon dioxide as the diluent, in an oxygen-modified air. Here is a summary of typical steps in data analysis [18].

1. Get the thermal signature (C_O, H_O, Q_F) of a pseudo fuel by summing up the energy terms of each fuel.
2. Reconstruct the explosibility diagram using the thermal signature for this pseudo fuel.
3. Get the global quenching heating ratio for this
4. Extract MOC and LOC from the flammability diagram.

The results of global indices are presented in Fig. 8.6.

Figure 8.6a shows the oxygen evolution history. The continuous drop of oxygen concentration shows that the sealing is well maintained in all 6 days. LOC stays constant, reflecting the fact that the fuel type is unchanged in the mixture. If the oxygen level drops below MOC, the mixture is non-flammable. If the oxygen drops below LOC, the mixture is non-explosive, or the flammable envelope shrinks to nil, which means the gases are no longer explosive. However, due to the energy trapped inside, the reaction continues locally since the temperature will expand the envelope and allow the reaction to continue. Due to insufficient oxygen supply, this reaction must be smoldering. In contrast, MOC is the ability to support flame propagation and a function of ambient oxygen and diluent, so it roughly follows oxygen measurement inside.

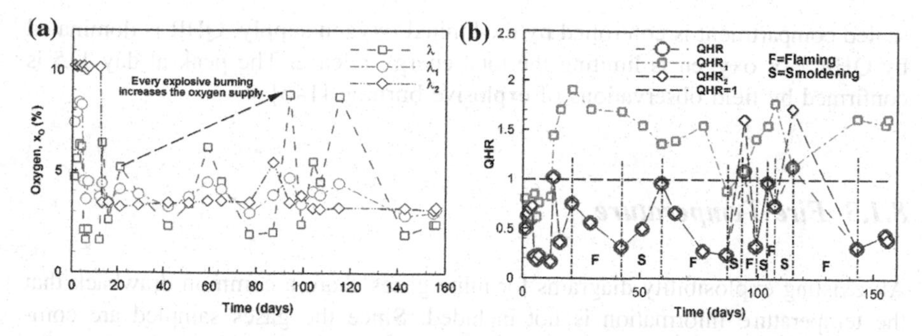

Fig. 8.7 The global indices for a control-failed mine fire. **a** Oxygen. **b** QHR

Figure 8.6b shows the fuel-dominated energy release (QHR1, ignitability) and the oxygen-dominated energy release (QHR2, explosibility). The minimum of the two (QHR, flammability) is the global explosibility index of the gas mixture. If QHR > 1, the gas mixture is flammable, as the point 3 and 4 shown in Fig. 8.6b. This is the most dangerous period with a high risk of explosion potential.

Generally, QHR index shows the flammable state of the mixture, while oxygen curves in Fig. 8.6a shows oxygen responses to the controlling efforts, which are better demonstrated in the next example.

A second mine fire came from India, where the original data are provided elsewhere [146]. Applying the global indices, the progress of this fire is better presented.

Following the procedures described above, Fig. 8.7 shows the oxygen evolution history at two sampling points. It clearly shows that there are several bursts of fire within this half-year period. If the oxygen level drops, it means the active burning inside and the positive pressure inside is preventing any oxygen intake. As the oxygen level drops below LOC, the reaction is so slow that there is insufficient pressure to prevent the oxygen from leaking-inside. Therefore, the oxygen level rises during this smoldering period, until the next explosive burning starts over again. Each explosive burning creates more oxygen leaks, so the burning intensity is higher and higher along the way.

Before the first explosive burning, LOC is about 10.2 %, which is mainly the LOC of methane. However, after the first explosive burning at day 13, the hydrogen production in the mixture pulls down LOC. At day 87, the hydrogen level is smaller and the LOC slightly rises. For most of the sealed burning time, LOC stays around 3.4, which is reflecting the major species production in a mine fire.

From Fig. 8.7a, we can find five peaks of oxygen measurement above LOC, signifying that there are five explosive burnings inside. The intensities and durations were increased overtime, showing an increased oxygen supply via the increased destroying power of explosive burnings.

From Fig. 8.7b, we can check the corresponding events represented as QHR peaks. Every smoldering stage will allow more oxygen to leak inside (increasing the explosibility), while every flaming stage will produce more fuels (increasing the ignitability). Smoldering and flaming are alternating over time. Since the fire in a sealed compartment is controlled by the limited oxygen supply, QHR is dominated by QHR2, or oxygen is limiting the total energy release. The peak at day 115 is confirmed by field observations of explosive burning [146].

8.1.3 Fire Temperature

All existing explosibility diagrams for mine gases share a common drawback that the temperature information is not included. Since the gases sampled are combustion results from the burning fire, the oxygen level is typically very low. By definition, if the oxygen level drops below LOC, the explosibility diagram will not be available [108]. There are several cases where the oxygen level really drops below LOC, showing that the impact of high temperature cannot be ignored. Assuming the oxygen measurement is the limiting oxygen concentration in a hot environment, we can find the scaled contribution of each component at the limit. Derived from these scaled contributions, the critical temperature to reach that oxygen level can be estimated. Remember, this approach only works for low oxygen levels (especially for those oxygen levels below LOC), which is caused strictly by explosive burning process, not valid for high oxygen levels, which are mainly caused by leaking from outside.

According to the general observation, the flame will extinguish if the ambient oxygen concentration drops below 11 %. However, all oxygen will be consumed in a furnace if ambient temperature is maintained high enough. Based on this reasoning, a higher ambient temperature will have a scaled contribution to the reaction. Starting from Eq. 7.23 and forcing R = 0, we have

$$\eta = \frac{C_O H_O}{1/x_L + Q_F - 1} \tag{8.5}$$

where x_L, C_O,H_O, Q_F are all mixture properties for the fuel mixture. However, η is the quenching potential of air, which is a simple function of temperature.

$$\eta(T) = 1.212 - 0.035 \cdot T^2 - 0.702 \cdot T \tag{8.6}$$

Solving these two equations, we can find the estimated reaction temperatures shown in Fig. 8.8. Note the assumption applies only to the low oxygen levels, which are the results of extinction without sufficient temperature. High levels of oxygen are the result of uncontrolled leaking, not burning, so they will generate some unrealistic low temperatures. Ignore these temperature drops, only the peaks are estimatable from the levels of low oxygen. The flue gas composition may tell more information about the combustion inside if the extinction criteria are considered properly. However, further experiments are needed to validate this theory.

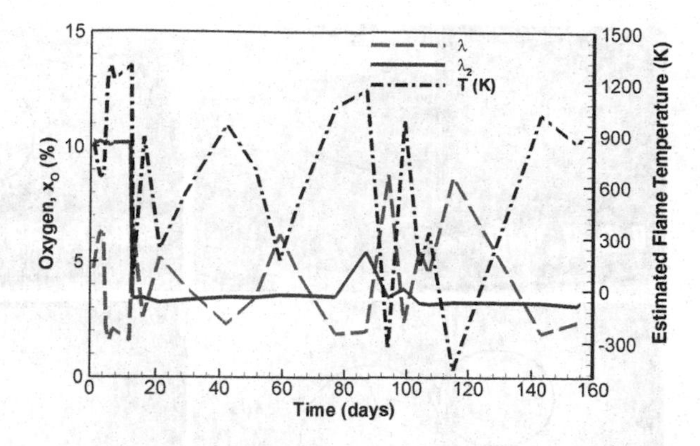

Fig. 8.8 Estimated flame temperatures as a result of oxygen depletion

8.2 Critical Behaviors in a Compartment Fire

8.2.1 Introduction to Backdraft

Backdraft is a critical fireground behavior, which endangers the lives of firefighters. It has many definitions, showing the complexity of the concept. According to NFPA, backdraft is the burning of heated gaseous products of combustion when oxygen is introduced into an environmental that has a depleted supply of oxygen due to fire. The burning often occurs with explosive force. The Institute of Fire Engineers defines backdraft as an explosion of greater or less degree, caused by the inrush of fresh air from any source or cause, into a burning building, in which combustion has been taking place in a shortage of air. Fire Research Station (FRS) defines backdrafts as the deflagration due to poor ventilation. "Limited ventilation can lead to a fire producing smoke gases containing significant proportions of unburnt gases. If these accumulate then the admission of air when an opening is made to the compartment can lead to a sudden deflagration, moving through the apartment and out of the opening. This deflagration is known as a "backdraft". According to the Swedish definition [149], a backdraft is the combustion of unburnt smoke gases, which can occur when air is introduced into the room where the oxygen content is significantly reduced due to the fire. Combustion can then occur more or less rapidly. A comparison of different definitions is provided in Table 8.4.

Here is our definition. A backdraft is a flame spread (deflagration) phenomenon, where fuel-rich (ignitable but non-explosive) mixture in a compartment is supplied with fresh air (ample with explosibility) and ignited, creating a fireball rushing out

Table 8.4 Definitions of backdraft

	NFPA	IFE	FRS	Sweden
Process	Burning	Explosion	Deflagration	Combustion
Cause	Oxygen supply	Air supply	Poor ventilation	Air supply
Results	Explosive force	–	Deflagration	Rapid combustion

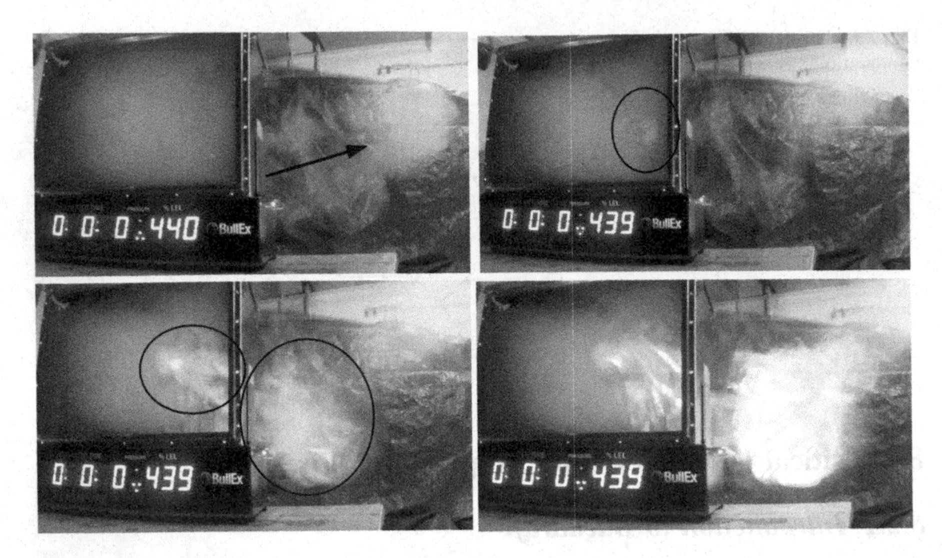

Fig. 8.9 The sequence of events for a backdraft produced on Bullex Flashfire simulator

of the ventilation. Since the ignition usually starts near the door, first draw-in the air, then push-out a fireball, it is called backdraft. Before the air supply begins, the mixture is ignitable, but not explosive. With the incoming air, the mixture is both ignitable and explosive, or flammable, creating a fireball rushing out of the ventilation.

With the Bullex® Flashfire simulator, we can reproduce a fireball mimicking a true progress of a backdraft. Here is a sequence of events for a backdraft. The simulation process is displayed in Fig. 8.9.

1. The chamber was filled with propane, up to a concentration of 2.2 % × 440 % = 9.7 %. The fuel vapor was visible with the addition of an aerosol. Then the door opened a cloud of fuel vapor rushed out.
2. The igniter near the door produced a spark, which ignited the premixed flame (bluish flame) near the door;
3. The premixed flame will rush out of the door, while the diffusion (yellowish) flame is still visible near the door. The fuel vapor inside is still too rich to get flame propagation throughout the space.
4. The burning fuel stream rush out of the door, forming a fireball.

From Fig. 8.9, we can see that a backdraft can occur when the fire is ventilation-controlled at a very early stage if the compartment is closed from the outset or if there are only limited openings. When the oxygen level drops the temperature in the room falls too. If the door to the room is then opened the smoke gases can ignite and cause a backdraft. When a backdraft occurs, usually the whole compartment is fully involved in flames.

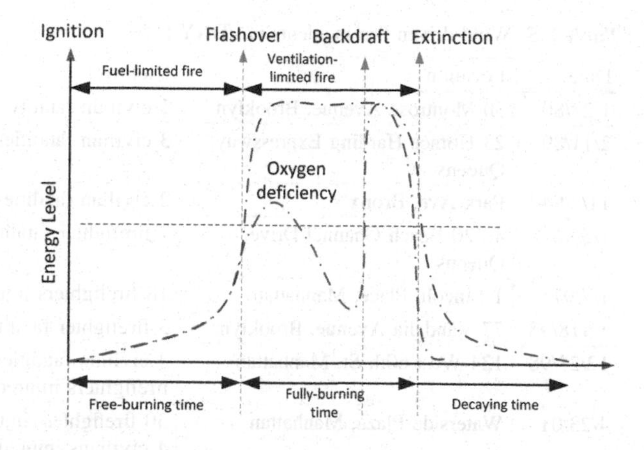

Fig. 8.10 Progress of backdraft and flashover in a compartment fire

Typically, a backdraft has the following features or steps:

- Unburnt fuel gases accumulate
- An oxygen-rich current comes in
- There is a well-mixed area of unburnt gases and air
- An ignition source ignites the gases in the well-mixed area
- A turbulent deflagration occurs in the room
- A fire ball is ejected out of the room.

The progress of a backdraft is demonstrated in Fig. 8.10. Flashover is also shown as a comparison. Initially, the fire is free burning given the fuel-limited burning status. As the burning area is growing, more fuel vapor will be released and flashover will follow. However, if the ventilation is limited, the transition to flashover is unfinished when the oxygen is depleted. The trapped energy will bring out more fuel vapors, forming a fuel-rich vapor. Suddenly, the air supply is resumed by breaking the sealed status. The ignitable but non-explosive mixture combined with air will suddenly burst into a fireball and rush out of the compartment. Without further fuel vapor supply, the fire will die down quickly. The hazard of a backdraft is typically reflected on the concentrated release of trapped energy, so it is a major hazard to firefighters in fireground operations.

8.2.2 *Blow-Torch Effect*

A similar phenomenon in fireground operations is called blowtorch effect. Unlike the sudden air supply through ventilation in a backdraft, this effect is caused by a sudden wind, leading to "a rapid developing fire that results from prevailing winds entering a fire-vented location of a structure, which pressurizes the interior, creating a deadly flow path of blowtorch effect flames and untenable temperatures when a secondary opening (vent point) is created" (Bowker's definition [150]).

Table 8.5 Wind-driven fire incidents in FDNY [150]

Date	Location	Victims	Stories	Fire floor
1/23/80	30 Montrose Avenue, Brooklyn	1 civilian fatality	16	11th
2/11/89	23 Horace Harding Expressway, Queens	3 civilian fatalities	16	14th
11/2/94	Park Ave, Bronx	2 civilian fatalities	20	18th
1/5/96	40-20 Beach Channel Drive, Queens	1 firefighter fatality	13	3rd
1/7/97	1 Lincoln Place, Manhattan	18 firefighters injured	42	28th
12/18/98	77 Vandalia Avenue, Brooklyn	3 firefighter fatalities	10	10th
12/23/98	124 West 60th St, Manhattan	4 civilian fatalities, 9 firefighters injured	51	19th
4/23/01	Waterside Plaza, Manhattan	30 firefighters injured, 4 civilians injured	37	24th
9/9/04	20 Confucius Place, Manhattan	12 firefighters burned	44	37th
1/26/06	40-20 Beach Channel Drive, Queens	3 firefighters burned	13	6th
2/26/06	20 Moshulu Parkway, Bronx	3 firefighters burned	41	24th and 25th
1/03/08	1700 Bedford Avenue, Brooklyn	1 firefighter fatality, 4 firefighters burned, 4 civilians injured	25	14th
3/28/08	Grand Avenue, Manhattan	1 civilian fatality, 45 injured	26	4th
4/2/08	Sutter Ave, Brooklyn	3 firefighters injured	22	5th

Wind is extremely dangerous in a high-rise fire, since the airflow is unstable above the atmospheric boundary layer. A sudden wind will bring the ample explosibility (in air) into the ignitable mixture (fuel), creating an immerse firewall which can disorient and trap unsuspecting firefighters instantaneously. New York Fire Department (FDNY) has observed many accidents involving blowtorch effects (Table 8.5).

Currently most research interests on blow-torch effect focus on the wind-driven effects and relating firefighting tactics [151]. Various curtains are proposed to block the air supply through open windows. Positive-pressure ventilation (PPV) are recommended to counter-balance the impact of wind. Madrzykowski [151] even advocates that no ventilation be attempted until after the fire has been knocked down in wind-driven fire conditions because of the danger of fire being pushed down on the interior crews. In a wind-driven fire with firefighter casualties, the Incident Command will be closely scrutinized and possibly be held responsible should anything happen.

However, the real danger is the smoke explosion behavior induced by a sudden air supply through the wind. In this perspective, a correct diagnosis of the imminent danger in a compartment fire is more important. Through an analysis of the smoke, we can diagnose the danger of an ignitable but non-explosive (fuel-rich) mixture, and take precautions against it. The flammable state of the smoke inside the compartment is the key to both backdraft and blowtorch effects.

8.2.3 Critical Conditions for Backdrafts and Blow-Torch Effects

Since the classical review by Croft [152], numerous researchers have studied this topic, however, due to the limitation of existing flammability theories, the critical backdraft conditions were represented as a fuel mass fraction of 0.10 [153], 0.16 [154], 0.098 [155], 0.0878 for natural ventilation and 0.1171 for mechanical ventilation [156], or the ratio between the volume fraction of combustible gases inside and the lower explosion limit of their mixture (still a fuel concentration) [157]. The reason for this diversity is that oxygen is not considered in the analyzing scheme, only fuel is considered. The impact of a variable oxygen level is completely ignored. With so many fuel criteria identified, it is difficult to apply them directly in guiding fireground operations. This fact limits the universality of one rule to cover all possible scenarios, including real mixtures from a compartment fire. Now with the new flammability and explosibility theory, it is possible to perform a fast analysis on the gas mixture and detect a potential smoke explosion through a gas analysis.

Traditionally, a ternary diagram is used to explain the occurrence of a backdraft, as shown in Fig. 8.11a. This ternary diagram is taken from Gottuk et al. [154] and has been adopted by a few other researchers [155, 156], while a diluted flammability diagram (Fig. 8.11b) can show a similar result.

Applying the concept of Heating/Quenching Ratio to the experimental data taken from Mao et al. [156], the flammable mixtures with and without backdrafts are clearly shown in Fig. 8.12. For those cases in which a backdraft occurred, the explosibility is increasing along with the air while the ignitability is decreasing due

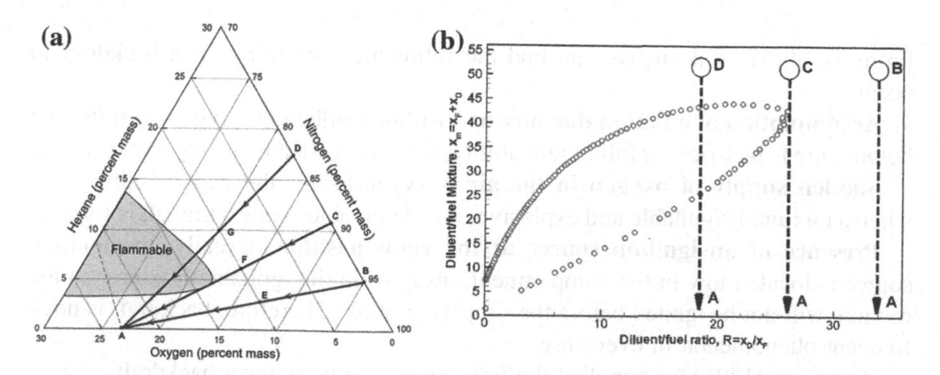

Fig. 8.11 Flammability diagrams for hexane in a backdraft (BC = no backdraft, D = backdraft occurred). **a** Ternary diagram [154]. **b** diluted flammability diagram

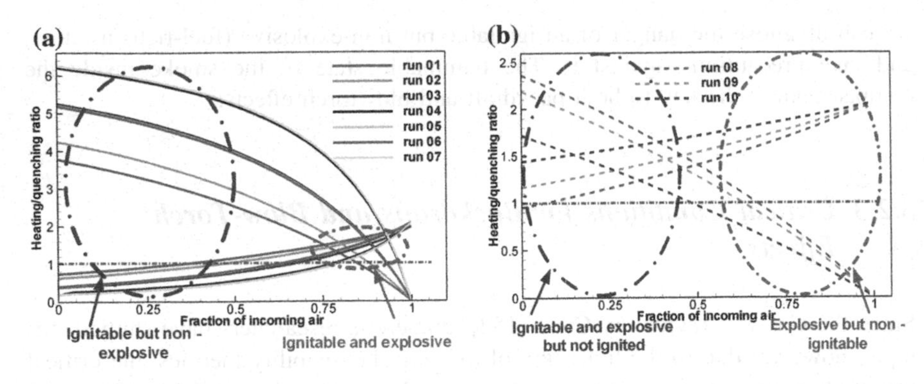

Fig. 8.12 The transition of states for an ignitable mixture based on experimental data by Mao et al. [156]. **a** Backdrafts occurred. **b** Without backdrafts

to dilution by air. If the non-ignitable state is reached (through dilution by air) before the mixture turns explosive (through oxygen addition), backdraft will never occur. This provides an effective means to detect the backdraft potential of any mixture. The essence of backdraft control is to reduce the ignitability (via purge by air) of the mixture before it gains sufficient explosibility (via oxygen intake) to be explosive (ignitable). The dangerous scenario is the case when both HQRs are above unity, which should be intervened before its occurrence.

For those cases with a backdraft occurred, the mixture is turning explosive along with the purging air supplying explosibility, while the ignitability is decreasing due to dilution by air. If the non-ignitable state is reached (through dilution by air) before the mixture is turning explosive (through oxygen addition), backdraft will never occur.

8.2.4 Controlling Strategies

From the above analysis, we can find the following conditions for a backdraft to occur:

Accumulation of a flammable mixture without sufficient oxygen supply. The hot mixture is fuel-rich, or full of ignitability, and oxygen-lean, lacking explosibility.

Sudden supply of oxygen in the air. Oxygen brings the explosibility. Only when a mixture is ignitable and explosive, it is flammable. Thus a fireball is formed.

Presence of an ignition source at the right position. Usually, the ignition source is located low in the compartment, away from the ignitable mixture, so the mixture will not be ignited before the supplying of air. Therefore, backdraft is not a frequent phenomenon in every fire.

Bengtssen [149] also supplied the following conditions for a backdraft.

Fuel's arrangement (or type of fuel). The lower the fuel's pyrolysis temperature, the easier to produce a fuel-rich mixture in the upper layer of a compartment fire. The higher up in a room the fuel is located, the more combustible pyrolysis product will accumulate there.

Initial opening's location and size. The lower down the opening is, the less pyrolysis products leak through the opening. If the opening is too small, the oxygen supply is small, so the fire may be smothered due to lack of oxygen and heat loss through the wall. If the opening is too large, the fire will transit smoothly into flashover without sufficient flammable smoke accumulation under the ceiling.

Insulation in the compartment. Reduced heat losses through the compartment wall will lead to a higher temperature rise in it. This temperature can stay for a long time, even if the fire is almost smothered by the limited ventilation.

Now, we can develop a qualitative control strategy for preventing a backdraft. The introduction of a diluent is to dilute the fuel stream, so the stream is non-ignitable upon leaking into air. This is a typical type II problem, or local application of diluting agents. Traditionally, only mist is considered as a suppression agent for preventing smoke explosions, because it is light and expanding upon evaporation. If vaporized, a drop of water will expand 1700 times, extremely well for diluting purpose or for local applications. Other light agents may have the same diluting effect as mist.

However, there is no quantitative theory to understand the role of mist on local application. Here based on the new theory in the diagrams shown in Fig. 8.12, the controlling strategies can be represented and compared in the following diagram (Fig. 8.13).

Curve A. The ignitability of the mixture (due to fuel) is always decreasing due to mixing with incoming agent. Curve B. The explosibility of the mixture (due to reduced oxygen) is increasing due to incoming air, which occurs as a result of opening the door for ventilation. Curve C. The explosibility of the mixture is controlled if the incoming agent (in air) has a same heating/quenching ratio (HQR), which is adjustable by adding water mist or some inert gases. Curve D. The explosibility of the mixture is completely suppressed by mixing with a pure inerting agent, nitrogen or water mist. The new flammability theory will provide a better comparison and leverage between different controlling strategies.

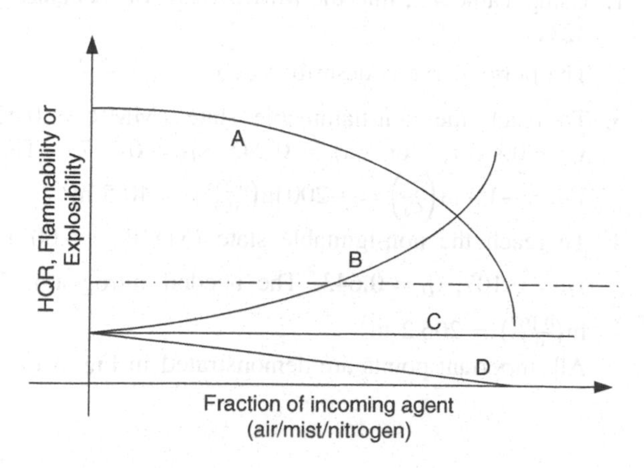

Fig. 8.13 Controlling strategies for preventing backdraft and smoke explosion

As long as the incoming mixture (diluent + air) will not bring up the explosibility of the mixture, the mixture with excess ignitability will be diluted and inerted eventually. This is the safety principle for preventing both backdrafts and blowtorch effects.

8.3 Problems and Solutions

Problem 8.1 A methane leak fills a 200 m^3 room until the methane concentration is 30 % by volume. Calculate how much nitrogen must be added to the room before air can be allowed in the space.

Solution:
Classical solution [64]:

The solution involves flammability diagrams and it can be shown that we need to reduce the methane concentration to 13 % in the mixture. The simple solution assume that as the N_2 is added, the initial mixture is ejected, and the amount of N_2 needed is

$$\frac{30-13}{30}\times 200 = 113\ \text{m}^3$$

However, there is generally no way to prevent mixing of the initial mixture with nitrogen before its ejection, so some of the nitrogen is lost. The most complicate solution considers this by assuming a "well-mixed" solution, leading to

$$C = C_0 \cdot exp\left(-\frac{V_N}{V}\right)$$

So $V_N = -V \cdot ln\left(\frac{C}{C_0}\right) = -200\, ln\left(\frac{13}{30}\right) = 167\ \text{m}^3$

New solution:

1. Using Table 4.2, find the MMR/MMF of methane, MMR = 6.005, MMR = 0. 4231
2. The purge curve is described as $x_m = \frac{(1+R)\cdot x_{F,0}}{R\cdot x_{F,0}+1-x_{D,0}}$
3. To reach the non-flammable state (MMR = 0.4231), we have $R = 0.72$, $x_m = 0.4231$, or $x_F = 0.245$, $x_D = 0.177$. The needed nitrogen is $V_N = -V\cdot \ln\left(\frac{C}{C_0}\right) = -200\ln\left(\frac{0.245}{0.3}\right) = 40.5\ \text{m}^3$
4. To reach the non-ignitable state (MMR = 6.005), $R = 6.005$, $x_m = 0.75$ or $x_F = 0.107$, $x_D = 0.643$. The needed nitrogen is $V_N = -V\cdot \ln\left(\frac{C}{C_0}\right) = -200 \ln\left(\frac{0.107}{0.3}\right) = 206.2\ \text{m}^3$

   All important points are demonstrated in Fig. 8.14.

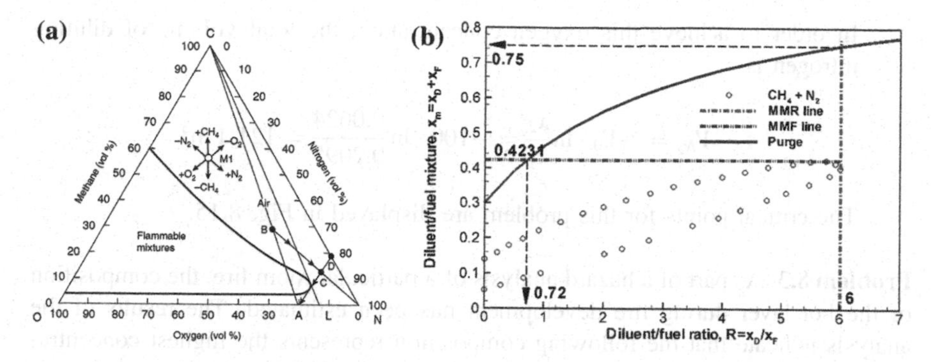

Fig. 8.14 Flammability diagrams {a Ternary [64] b Diluted} for purging

Problem 8.2 There is a potential CO leakage in a 100 m^3 room, how much diluent (N_2) is needed to fully inert this compartment against any CO leakage? Without an experimental flammability diagram, redo your estimations based on theoretical flammability diagrams.

Solution:

This problem can be solved in two ways, using experimental data and using theory.

(a) Using experimental data, the minimum inertion concentration is directly read from the standard flammability diagram as MIC = 0.575 for Nitrogen. So the initial oxygen concentration is 0.2095, the target oxygen concentration is 0.2095*(1 − 0.575 − 0.575/4.079) = 0.0595. Or MOC is estimated as $MOC = 0.2095 \cdot (1 - x_{LU}) = 0.2095 \times (1 - 0.709) = 0.061$.
In order to achieve this oxygen concentration, the total volume of diluting nitrogen is

$$V_{N_2} = -V_0 \cdot \ln\frac{x_2}{x_1} = -100 \cdot \ln\frac{0.0595}{0.2095} = 125.9\,\text{m}^3$$

(b) without experimental data, the inertion point for nitrogen is determined as

$$R_{LU} = \frac{C_O H_O \lambda - \lambda Q_F - C_O Q_m}{\lambda Q_D} = \frac{0.2095 \times 0.5 \times 15.59 - 0.2095 \times 0.8 - 0.5 \times 1}{0.2095 \times 0.992} = 4.64$$

$$x_{LU} = \frac{Q_m}{Q_m + \frac{C_O H_O}{1+R_{LU}} - \frac{Q_F}{1+R_{LU}} - \frac{Q_D R_{LU}}{1+R_{LU}}} = \frac{1}{1 + \frac{0.5\times15.59}{1+4.64} - \frac{0.8}{1+4.64} - \frac{0.992\times4.64}{1+4.64}} = 0.702$$

Then the critical concentration is determined as

$$MIC = \frac{R_{LU}}{1+R_{LU}} \cdot x_{LU} = \frac{4.64}{1+4.64} \cdot 0.702 = 0.578$$

$$MOC = 0.2095 \cdot (1 - x_{LU}) = 0.2095 \times (1 - 0.702) = 0.0624$$

In order to achieve this oxygen concentration, the total volume of diluting nitrogen is

$$V_{N_2} = -V_0 \cdot \ln\frac{x_2}{x_1} = -100 \cdot \ln\frac{0.0624}{0.2095} = 121.1\,\mathrm{m}^3$$

The critical points for this problem are displayed in Fig. 8.15.

Problem 8.3 As part of a hazard analysis of a particular room fire, the composition of the hot layer during fire development has been estimated. The results of the analysis indicate that the following composition represents the highest concentration of fuel gases expected:

Hot layer temperature = 700 K, 10 % methane, in the form of CH_4, 2 % CO, 1 % H_2, 15 % CO_2, 2 % O_2, 70 % N_2. Cold layer temperature = 300 K, 20 % O_2, 79 % N_2. Will this hot layer burn? [79]

Solution:

Step 1. Find the thermal signature of the fuel mixture through weight-averaging in spreadsheets.

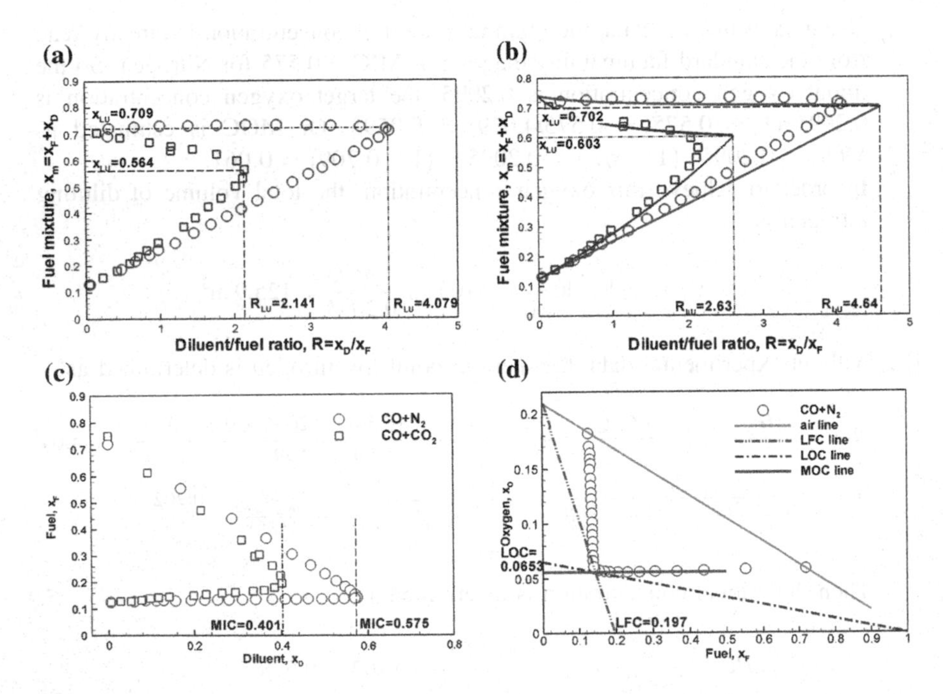

Fig. 8.15 Flammability diagrams for CO safety. **a** Experimental diluted diagram. **b** Theoretical diluted diagram. **c** Standard flammability diagram. **d** Explosive triangle diagram

	CO	CH_4	H_2	Mixture
C_O	0.50	2.00	0.50	1.65
x_L	12.50	5.00	4.00	
x_U	74.00	15.00	75.00	
Q_F	0.80	13.81	3.51	11.01
H_O	15.59	16.40	55.02	19.25
Fraction	0.15	0.77	0.08	

Step 2. Find the thermal signature of the mixture

$$FHP = C_O H_O x_F = 1.65 \times 19.25 \times 0.13 = 4.13$$
$$OHP = H_O x_O = 19.25 \times 0.02 = 0.385$$
$$QP = \sum Q_{F,i} \cdot x_i = 1.05 \times 0.02 + 15.59 \times 0.02 + 16.40 \times 0.1$$
$$+ 55.02 \times 0.01 + 1.75 \times 0.15 + 0.992 \times 0.7 = 2.41$$

Step 3. Find the temperature correction factor for the hot layer.

$$\eta(T) = \frac{E_i}{E_{air}} = 1.212 - 0.035 \cdot T^2 - 0.702 \cdot T$$
$$= 1.212 - 0.035 \times 0.7^2 - 0.702 \times 0.7 = 0.70345$$

This hot layer temperature will affect the quenching potential of the mixture

$$QP = 2.41 \times 0.70345 = 1.695$$

Therefore, the mixture has

$$HQR_1 = \frac{FHP}{QP \times \eta(T)} = \frac{4.13}{2.41 \times 0.70345} = 2.44$$
$$HQR_2 = \frac{OHP}{QP \times \eta(T)} = \frac{0.385}{2.41 \times 0.70345} = 0.23$$

This hot layer is ignitable ($HQR_1 > 1$), but non-explosive ($HQR_2 < 1$).

Step 4. If mixing with cold air, HQR_1 is decreasing, while HQR_2 is increasing

$$HQR_1 = \frac{4.13 \times (1 - x_{air})}{2.41 \times 0.70345 \times (1 - x_{air}) + 1 \cdot x_{air}}$$
$$HQR_2 = \frac{0.385 \times (1 - x_{air}) + 0.2095 \cdot x_{air} \cdot H_O}{2.41 \times 0.70345 \times (1 - x_{air}) + 1 \cdot x_{air}}$$

These two curves are displayed in the diagram below.

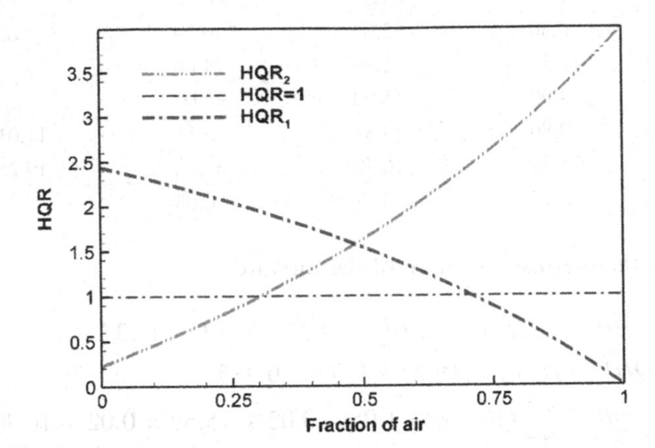

This diagram shows that if diluted by cold air, the mixture has a potential to ignite and burn between the air fraction of 0.28 and 0.72. Before that range, the mixture is ignitable, but non-explosive. After that, the mixture is explosive and non-ignitable.

Chapter 9
Summary and Conclusions

Instead of focusing on the delicate energy balance at critical limits, the conservation of energy by mass transfer is explored in this monograph. More fundamental concepts, such as ignitability and explosibility, are given new meanings using oxygen thresholds. Here is a review of basic assumptions and what can be achieved using the thermal balance method. It provides a systematic approach or an engineering approach toward common safety problems on mixture flammability.

9.1 Review of Basic Assumptions

9.1.1 Assumption on Constant Flame Temperature

In order to apply the energy balance, one necessary assumption is the constant flame temperature at limits. Similarly, one of the premises to apply Le Chatelier's rule is the assumption on constant flame temperature [66]. The flame is only propagated (or ignited, or burnt) if a certain minimum temperature can be maintained in the flame boundary, which is approximately constant for a given combustible. Here we will check the history of this assumption by previous researchers.

The idea of using flame temperature for predicting gas limits is not new, it was assumed in 1914 [158] that the ignition temperature of a gas mixture at concentrations similar to those present in a lower-limit mixture does not vary much with the concentration of the inflammable gas. This assumption has not been proved to hold generally, but it has been shown to hold for hydrogen-oxygen mixtures by Dixon and Crofts [159], while Tizard and Pie [160] believe that the ignition temperature changes little throughout the whole range for Heptane-air. It has also been shown [44] that at the lower limit in ammonia-air and -oxygen mixtures the theoretical flame temperature remains practically constant for a range extending from the ordinary temperature to 450 °C.

Egerton [50] found that there is no synergistic effect in flame propagation, and the adiabatic flame temperature at the lean flammability limit remains approximately constant in hydrocarbon-air mixtures. His observation supports the thermal

T. Ma, *Ignitability and Explosibility of Gases and Vapors*,
DOI 10.1007/978-1-4939-2665-7_9

view in this work that the flammability is a result of a thermal balance between heat release and heat absorption, with other heat losses playing a minor role on flame propagation. Egerton is a pioneer on the chain-reaction theory of combustion, who insists the critical flame temperature on flame propagation. Whether combustion takes place or not is dependent upon the temperature of the flame being sufficient to maintain a certain boundary flame temperature. This threshold provides a certain concentration of active radicals in the boundary region and inflammations produced in the unburnt gas mainly from the radical derived from the flame [35]. However, White [44] showed that there was no relation between limits for propagation and flame ignition temperature as originally projected.

In reality, there is no such a constant flame temperature to follow. Dixon and Coward [161] showed that the ignition-temperatures of the paraffin hydrocarbons appear to fall slightly as the series (C_nH_{2n}+2) is ascended. In attempting to use this equation for the calculation of the quantities of the paraffin hydrocarbons necessary, one is faced with the difficulty that their ignition-temperature are very ill-defined [6].

However, if Burgess and Wheeler's relations holds true, the critical adiabatic flame temperature should be nearly constant [5]. Zabetakis [43] used 1300 °C as the threshold for flammability. It has also been common knowledge for a long time that the computed (equilibrium) adiabatic flame temperature for lean-limit mixtures in air cluster near 1500 or 1600 K for many hydrocarbon-type fuels [104]. Mashuga and Crowl [162] used a 1200 K threshold and obtained reasonable estimates of the flammable regions for the gases methane and ethylene. Hertzberg et al. [163] showed that methane flammability could be predicted using CAFT isotherms in the 1500–1600 K range. Melhem [41] used a 1000 K threshold to predict flammability limits. Beyler [164] reviewed the history of using 1600 K as the critical threshold for predicting fire extinction. In a recent research [165], a CAFT threshold temperature of 1200 K was used to predict flammability limits of vapor mixtures.

The real flammability limit temperatures are not constant. However, there is no published database for determining flammability by flame temperature. Some algorithms for predicting the limit temperature, such as Mathieu's work [166], were proposed. However, the safety problem does not require a precision on critical limits, so it provides the arena for engineering methods in this monograph.

9.1.2 Assumption on Oxygen Calorimetry

The role of oxygen has two meanings on flammability theory. Firstly, it means a conversion from chemistry to energy (physics). White [44] noticed that the amount of oxygen available for the combustion of a vapor in its limit mixture bore a fairly constant ratio to the amount required for the perfect combustion of 1 molecule of the vapor. That means the "oxygen using capacity" is roughly constant. Egerton and Powling [35] confirmed this finding, "the heat of combustion per mole of the mixture" is most important in governing the limit [5]. Britton [48] stresses the universal role of oxygen in flammability theory, and used the heat of oxidization in his series of correlations. The ratios of the lower limits of the individual constituents to the amount of oxygen required for theoretical perfect combustion are about the same. Huggett [90] further generalized this principle for solid fuels, and called it oxygen calorimetry, which later became the foundation of heat release measurement. Under this reasoning, more correlations based on stoichiometric oxygen number were proposed than on the heat of combustion [17]. Britton [59] used this oxygen calorimetry as the threshold for defining explosive materials.

Secondly, the availability of oxygen limits the heat release at the upper limits. As Thornton [89] first proposed, the upper limit bears a direct relation to the amount of oxygen needed for prefect combustion (theoretical complete combustion), which is called Thornton's rule. This rule makes the energy conservation at upper limits possible, which is also vital to the success of the thermal balance method. Explosibility is based on the delicate balance between oxygen-based heat release and mass-based heat absorption. Therefore, Thornton's rule sets the foundation of explosibility. Jones and Kennedy [107] discussed the prevention of gas explosion by controlling oxygen concentration. The diluted flammability diagram was proposed to guide the inerting operations. They found that carbon dioxide is more effective than nitrogen as a diluent, so the concept of MOC and flammability was implicitly assumed. Their method is still a major means against potential explosions in a confined space (type I problem).

9.2 What Is New?

As a summary, we need to find out what is new in this monograph. Frankly, there is nothing first proposed in this work. Major concepts first appeared elsewhere. For example, Crowl and colleagues [66, 109, 162, 167] have utilized the concepts of ISOC and OSFC in protecting storage tanks, while Schröder and his colleagues [70, 78, 127, 168, 169] have identified critical points and lines in a ternary diagram. What is missing in the past is a rigorous theory from fundamental principles. Here a framework of theory is established upon the thermal balance, allowing us to check all flammability-related concepts and methods. So old concepts are given new meanings under the principle of energy conservation. In summary, this monograph is built upon one method, two problems, three concepts and four diagrams, with five applications. Let us go over them one by one.

9.2.1 Thermal Balance Method

The concept of using CAFT to find the mixture limits is not new. Hansel [170] has already adopted this concept in solving mixture problems using commercial software while Mashuga et al. [162] tried to develop flammability diagrams using critical temperatures. What is new here is a binary view on each species in the mixture [15]. All species has a mass, with a non-zero quenching potential, while fuel or oxygen can release energy, with LFL controlled by fuel and UFL controlled by oxygen. This binary treatment makes the contribution of each species cumulative and additive, thus making a hand-calculation possible. The primary utility of this method is to find the flammable limits of a flammable mixture. Since the estimated results are highly dependent on the accuracy of the inputs, any errors introduced by initial measurements will be passed on to the estimated results. For some flammability data, the published LFL may be small in value. Caution should be taken in using these data for predictions. This method is in essence a conversion scheme, converting one limit condition to another, which has been confirmed by flammability diagrams.

In fact, all estimation methods are dependent on the assumption on flame temperature or flame structure explicitly or implicitly. Hydrogen changes the flame thickness through its diffusivity, thus decreasing the flame temperature. Halon 1301 changes the flame temperature through its scavenging effect on reacting ions. Both will be difficult to predict using this theory. This limitation applies to other estimation methods, including LCR and Beyler's method, because all estimation methods are based on the assumption of a constant CAFT. This method is better than other methods because it considers the contributions from the quenching potentials of fuel/oxygen/diluents, which are not covered in other works. Though equivalent with Le Chatelier's Rule mathematically, a comparison is provided in Table 9.1.

The utility of the thermal balance method has grown beyond mixture flammability. In addition to mixture flammability estimation, the thermal balance method provides additional functions:

1. A temperature dependence replacing modified Burgess-Wheeler's law;
2. Theoretical flammability diagrams based on limited inputs;
3. Critical points and lines on ignitability, flammability and explosibility.
4. A graphical representation of controlling variables and safe operation routes.

Table 9.1 Comparison of the thermal balance method with Le Chatelier's Rule

	Thermal balance method	Le Chatelier's Rule
Limitation on components	No limitation	Fuels only
Temperature dependence	Mixture enthalpy	Background air enthalpy only (modified Burgess-Wheeler's law)
Treating diluents	Using quenching potentials	Using diluted flammability diagram or ISO10156 method
Nature	Derived from fundamental principles	Empirical

9.2.2 Two Types of Problems

This work also solves two types of flammability problem. Type I problem is a static inertion problem, with flammable mixture in a confined space. The purpose of inertion is to control the explosibility in the mixture, where the oxygen level can be controlled in a confined space. Total flooding the space is a typical type I problem. For type I problem, we have MOC and LOC, representing flammability and explosibility respectively. MOC is required for supporting the flame propagation, while LOC is required for supporting explosive burning. **LOC** defines the minimal oxygen requirement to fully consume the fuel, so it is a fuel property. Each fuel has its unique LOC with a specific diluent. **MOC** defines the oxygen level to support flame propagation, which is a process property. MOC only applies to the condition of the inertion point (or nose point), where the explosibility (LOC) line and the ignitability (LFC) line cross each other. MOC is a concept in flammability, while LOC is a concept in explosibility. It is difficult to derive a simple expression for MOC, since the energy conservation for flame propagation is too delicate to develop into a mathematical expression.

If the oxygen level drops to MOC, we defined $\lambda < \lambda_1$ as "non-flammable" (since oxygen is lean), or simply "inerted" (since the role of diluent is both diluting and quenching). LOC provides the necessary oxygen to fully consume the fuel, while it can be fully determined from the thermal properties of the fuel. Or LOC is the total oxygen requirement to match fuel (LEL), so it is more fundamental. If $\lambda < \lambda_2$, the flammable envelope is shrinking to nil, so the mixture is non-explosive or fully inerted. Here "inerted" is a good expression, since the role of diluent is not only diluting (causing oxygen depletion), but also heat absorption.

In contrast, type II problem is a dynamic dilution problem with flammable streams leaking into ambient air. The purpose of dilution is to control the ignitability of the fuel stream, which is released into the air of fixed explosibility. Local application of a suppressant is a typical type II problem. Since the background air in the open cannot be controlled, more agents are needed to dilute the fuel so the ignition is not possible in air. For type II problem, we have MFC and LFC, representing flammability and Ignitability respectively. Limiting Fuel Concentration (LFC) is used to control the ignitability of the fuel stream. Analogous to LOC, **LFC** is the critical fuel fraction, which does not support ignition, even in air. **MFC** is the critical fuel fraction when air is involved at the inertion point or the nose point. Therefore, MFC is a concept in flammability, while LFC is a concept in ignitability. If $\lambda_2 = 0.2095$ is reached, the diluted flammable envelope is tilted outside the effective domain, or the background oxygen is insufficient to support ignition. Then no ignition is possible. So we label it as "non-ignitable" (since fuel is too lean) or simply "diluted". Previously, we say the fuel is inerted by the diluent, if it cannot support ignition of any kind. Here "diluted" is a preferred word to stress the lean state of the fuel mixture and the volume displacement by a diluent.

A side-by-side comparison between two types of problem is shown in Table 9.2.

For an inerting problem (type I problem), the role of a diluent is to reduce the concentration of oxygen in the background, or reducing the explosibility of air. For a

Table 9.2 Summary of two types of problems on flammability

	Type I problem	Type II problem
Typical scenario	Sealed underground coal fires or filling a liquid fuel tank	Leakage of a flammable/diluent mixture into air or emptying a liquid fuel tank
Typical question	Is the mixture self-flammable?	Is the mixture ignitable in air?
Fuel	Confined space	Open/confined space
Operation	Purge/dilution	Dilution
Suppression mode	Total flooding	Local application/total flooding
Safety goal	First control explosibility (oxygen), then control ignitability (fuel)	Control ignitability (fuel)
Role of diluent	First, inerting air by diluting the oxygen, later inerting mixture by diluting the fuel	Inerting fuel by diluting the fuel
Target	Control the flammability (oxygen) and explosibility (fuel) of the mixture	Control the ignitability (fuel) of the mixture
Target of inertion	With fuel, $x_O < \lambda_1$ (non-flammable)	$x_O = 0.2095 < \lambda_2$ (non-ignitable)
	Without fuel, $x_O < \lambda_2$ (non-explosive/ignitable)	
Fuel	–	Without oxygen, $x_F < \beta_2$ (non-ignitable)
		With oxygen, $x_F < \beta_1$ (non-flammable)
Diluent	With fuel, $x_D > MIC$ (non-flammable)	With oxygen, $x_D > MIC$ (non-flammable)
	Without fuel, $x_D > 1 - 4.773\lambda_2$ (non-explosive/ignitable)	Without oxygen, $x_D > LDC = 1 - \beta_2$ (non-ignitable)
Critical parameters	$MMF = x_F + x_D > 1 - 4.773 \cdot \lambda_1$	$MMR = x_D/x_F > R_{LU}$
	$MMR = x_D/x_F > R_{LU}$	

dilution problem (type II problem), the role of a diluent is to reduce the concentration of fuel in the fuel stream, or reducing the ignitability of fuel. Combined together, the flammability of the mixture is determined. They have different working mechanisms and controlled by different flammability diagrams as shown below.

9.2.3 Three Concepts

Though ignitability and explosibility are not new concepts, they are poorly defined in the past. Here they are reused to label the property of fuel and air respectively. A demonstration of three critical concepts is supplied in Fig. 9.1. By analyzing the physical meaning in a flammability diagram, it is found that ignitability is

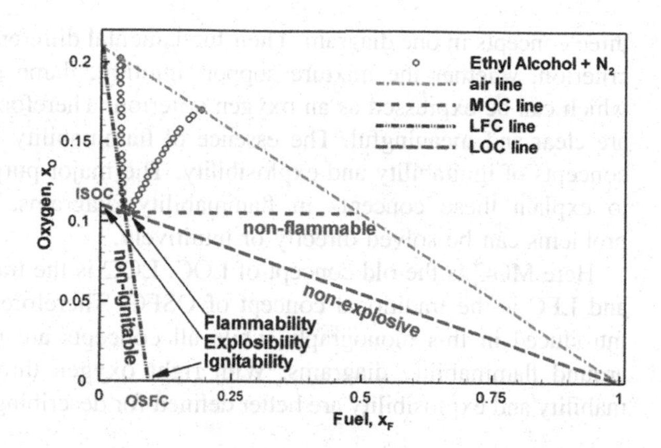

Fig. 9.1 Three concepts are properly demonstrated in an explosive triangle diagram

fundamental to fuel, while explosibility is fundamental to air. Flammability is a special case (at the inertion point) on both LOC and LFC lines. If this inertion point can be determined by reading experimental flammable envelope, all critical operation parameters can be derived, giving it a pivotal position in any flammability diagram. By establishing a database (like Table 4.2) on this inertion point, it is possible to derive any dilution requirement without the help of a flammability diagram.

A comparison of these three concepts is provided in Table 9.3 to show the utility of these concepts in solving flammability problems. Figure 9.1 demonstrates these

Table 9.3 Summary of three critical concepts in flammability diagrams

	Ignitability	Flammability	Explosibility
Property	Material property	Process property	Material property
Associated with	Fuel	Fuel/oxygen	Oxygen
Safe status	Non-ignitable (fuel-lean, fully diluted)	Non-flammable	Non-explosive (oxygen-lean, fully inerted)
Physical meaning	Background oxygen in air is insufficient to support ignition	Insufficient oxygen in air to support flame propagation	Insufficient oxygen in air to support explosive burning
Flammable envelope	Flammable envelope tilts out of the domain by dilution	–	Flammable envelope shrinks to nil by inertion
Problems	Type I/II	Type I	Type I/II
Fuel	$x_F < \beta_2$	$x_F < \beta_1$	$x_F = 1 < \beta_2$, impossible
Oxygen	$x_O = 0.2095 < \lambda_2$	$x_O < \lambda_1$	$x_O < \lambda_2$
Concepts	LFC/LDC	MMF/MIR/MMR/ MOC/MFC/MIC	LOC/LIC
Major role of diluent	Inerting fuel by diluting the fuel	Inerting air by diluting both the oxygen and the fuel	Inerting air by diluting the oxygen

three concepts in one diagram. Their fundamental difference come from the ignition criterion, whether the mixture support ignition, flame propagation or explosion, which can be expressed as an oxygen criterion. Therefore, their physical meanings are clear and meaningful. The essence of flammability lies on more fundamental concepts of ignitability and explosibility. The major purpose of this monograph is to explain these concepts in flammability diagrams, so all gas-safety related problems can be solved directly or intuitively.

Here MOC is the old concept of LOC, LOC is the traditional concept of ISOC, and LFC is the traditional concept of OSFC. Therefore, there is no new concept introduced in this monograph, while all concepts are redefined and reorganized around flammability diagrams. With right oxygen thresholds, ignitability, flammability and explosibility are better defined for describing critical points in Fig. 9.1.

9.2.4 Four Diagrams

Finally, we have four types of flammability diagrams discussed in detail. Given the importance of critical points in determining the flammable state and purge/dilution requirements, we can demonstrate the purge or dilution routes in any of the four flammability diagrams. All diagrams are equivalent and interchangeable with a simple conversion scheme. Using which diagram depends on the operation target and familiarity. Here we have a comparison of four diagrams summarized in Table 9.4.

These four diagrams have different priorities in presenting data. Standard diagrams are used frequently for presenting experimental data, while ternary diagrams are usually used to demonstrate the dilution and purge operations. A diluted diagram is most fundamental in presenting the inertion point, while an explosive triangle is comprehensive to demonstrate major critical points. Failing to compare all four diagrams is a major reason of conflicting views on various critical points. This monograph tries to unify these concepts with the help of all four diagrams.

Table 9.4 Comparison of flammability diagrams

	Standard diagram	Diluted diagram	Explosive triangle	Ternary diagram
Inputs	x_D, x_F	R, x_m	x_F, x_O	x_D, x_F, x_O
Major outputs	MIC	MMR, MMF	LOC, LFC, LDC/LIC, MOC, MFC, MIC	LOC, LFC, LDC/LIC MOC, MFC, MIC
Advantages	Read dilution requirement directly	Fundamental parameters; easy to derive other parameters	Clear definitions of LOC and LFC	Reading data directly; clear definitions of LOC and LFC
Disadvantages	Difficult to find LOC	Need a conversion to get x_D, x_F, x_O	Need a conversion to get MIC	Difficult to apply analytical solutions

Though the method outlined in this monograph is powerful enough to solve all flammability-related problems, five typical problems are solved in this work. They are:

1. The analysis and diagnosis of combustion products from a sealed mine fire;
2. The compartment fire diagnosis for backdrafts and blow-torch effects;
3. The factors affecting the flammability of clean-combustion technologies;
4. The critical flammability ratio for refrigerant mixtures;
5. Safe operations before filling or emptying a fuel tank.

9.3 Why a Comprehensive Flammability Theory Is Difficult to Establish?

Here comes the final question, why a flammability theory is difficult to establish? After 200 years of experimental and theoretical work, we are still living on islands of knowledge without a coherent theory on flammabilty.

One reason deals with the complexicity of flame propagation. Radiative heat loss [171] determines the flame speed, whereas the ambient pressure [172] makes a contribution. The original definition of flammability, to support flame propagation indefinitely, is ambiguous on the flame speed, so there is some uncertainty on critical threshold. The flame propagation criterion is simple in experiments, but difficult to quantify in numerical or theoretical analysis. Numerical experiments, such as [121] and [172], can have different results if the threshold is changed, as shown in Fig. 7.3. This puts the prediction of mixture flammability in a delicate and sensitive position. If one of the ambient parameters is changed, the results can be sharply different. In contrast, mass transfer is dominating the heat transfer process. Flame temperature change primarily as a result of mixing. Only when the temperature is approaching the critical value, radiative loss fraction becomes significant. The methodology in this monograph is focusing on the mass transfer side instead of the heat transfer side of a flammability problem.

Another reason is that flammability is a process concept or a mixture property, which relies heavily on local oxygen and diluent to support the flame propagation process. Flame propagation always depends on a series of environmental factors, such as the direction of flame propagation, the design, diameter, and length of the test apparatus, the temperature and pressure of the mixture at the time of ignition, the percentage of water vapor present, and indirectly by the source of ignition. Therefore, flammability is not a fundamental material property. It was chosen mainly because the ignition criterion is easy to follow in flammability tests, while an ignitability or explosibility test is too ideal to perform. In addition, the MFC line and the MOC line are used as dividing lines, both are not fundamental as the LOC line or the LFC line. Therefore, flammability is not fundamental to that fuel. Only because flammable limits and explosive limits are close to each other for a pure fuel burning in air, and explosive limits and ignitable limits are equivalent in a confined space (type I problems), it was assumed flammability and explosiblity are interchangeable, giving the flammability a dominant position for the past two centuries.

Zabetakis [43] noticed that the modified Burgess-Wheeler's law cannot predict the temperature dependence correctly. The surface reason is that the contribution of fuel is not included in the extrapolation scheme, while the core reason is that Burgess-Wheeler's argument is based on the concept of ignitability (of fuel), while most data are taken under flammability tests (of mixture). Without knowing the difference between flammability and ignitability, most energy conservations were biased towards fuel, further complicating the flammability problem where oxygen plays a vital role. Failing to differentiate the concepts between explosibility and ignitability is the major reason that the energy principle does not work properly in almost all previous work.

In contrast, explosibility and ignitability are more fundamental and abstract concepts. Explosive limit is the limiting oxygen concentration (LOC) for supporting explosive burning, while ignitable limit is the limiting fuel concentration (LFC) for supporting ignition initiation. Explosibility is a simple conversion from oxygen to fuel and the capability of background air to support flame propagation, being more fundamental to the background air. Ignitability is the capability of the fuel in the stream to support ignition, fundamental to that fuel. Theoretically and idealistically, you need to remove fuel to find the explosibility of the background air, or remove oxygen to find the ignitability of the fuel stream. They are not easy to realize in any experimental setup. However, Ishizuka and Tsuji [113] and Simmons et al. [114] did measure the explosibility and ignitability in counter-diffusion flames. Instead, the critical ignitability (LFC) line and the critical explosibility (LOC) line can only be extrapolated from the flammability diagram, and they decide the position of the inertion point, or the flammability. Flammability is a delicate balance between explosibility and ignitability, not a fundamental property of that fuel. A summary of major misunderstandings is provided in Table 9.5.

There is nothing new under the sun. For the fundamental concepts of explosibility (ISOC/LOC) and ignitability (OSFC/LFC), Ishizuka and Tsuji [113] and Simmons et al. [114] measured them, Crowl and colleagues [109] extrapolated

Table 9.5 Misunderstandings hindering a coherent flammability theory

	Traditional views [70, 109]	Current views
Energy conservation	Fuel-based energy release: $x_L \cdot \Delta H_C = K$	Oxygen-based energy release: $\lambda_2 \cdot \frac{\Delta H_C}{C_O} = K$
Limiting oxygen concentration	Flammability: $\lambda_1 = C_O \cdot x_L$	Explosibility: $\lambda_2 = \frac{C_O \cdot x_L}{1-x_L}$
Role of fuel	Out-of-service fuel concentration (OSFC): $x_{OSFC} = \frac{\lambda_2}{C_O \cdot (1-\lambda_2/x_0)}$	Ignitability: $\beta_2 = \frac{1}{1+R_{LU}}$
Role of oxygen	In-service oxygen concentration (ISOC): $x_{ISOC} = \frac{C_O \cdot x_L}{1-x_L}$	Explosibility: $\lambda_2 = \frac{C_O \cdot x_L}{1-x_L}$

them in a ternary diagram. Schröder and his colleagues [70] utilized these concepts implicitly in developing ISO10156 methods. However, they all failed to relate the ignition criteria to their findings. These ignition criteria are closely related to a fuel-based energy release or an oxygen threshold. These concepts are not following the classical definition of flammability, to propagate flame indefinitely. They are unified in this monograph.

In order to utilize the original definitions for ignitability, flammability and explosibility, a framework of mathematical treatment is needed. Thermal balance method in this work is serving as a powerful tool to derive the ideal cases (explosibility and ignitability) and the special case (flammability). With clearly defined oxygen thresholds, ignitability, flammability and explosibility are also properly defined. Utilities of various flammability diagrams are demonstrated in safe operations (dilution and purge) or suppression modes (total flooding and local application) on flammable gases and vapors. This is the major mission of this monograph and mission is accomplished.

References

1. S.H. Davy, On the fire-damp of coal mines, and on methods of lighting the mines so as to prevent its explosion. Philos. Trans. R. Soc. Lond. **106**, 1 (1816)
2. J.M. Kuchta, *Investigation of Fire and Explosion Accidents in the Chemical, Mining, and Fuel-related Industries: A Manual* (US Department of the Interior, Bureau of Mines, Pittsburgh, 1985)
3. L.G. Britton, Two hundred years of flammable limits. Process Saf. Prog. **21**, 1–11 (2002)
4. H. Coward, G. Jones, *Limits of Flammability of Gases and Vapors*. Bulletin, vol. 503 (Pittsburgh: US Department of the Interior, Bureau of Mines, 1952)
5. A.C. Egerton, Limits of inflammability, in *Fourth Symposium on Combustion* (Williams and Wilkins, Baltimore, 1953)
6. M.J. Burgess, R.V. Wheeler, The lower limit of inflammation of mixtures of the paraffin hydrocarbons with air. J. Chem. Soc. **99**, 2013–2030 (1911)
7. I. Glassman, *Combustion* (Princeton University, Princeton, 1996)
8. L. Lovachev et al., Flammability limits: an invited review. Combust. Flame **20**(2), 259–289 (1973)
9. G. Jones, Inflammation limits and their practical application in hazardous industrial operations. Chem. Rev. **22**(1), 1–26 (1938)
10. R.W. Van Dolah et al., Ignition or the flame-initiating process. Fire Technol. **1**(1), 32–42 (1965)
11. I.A. Zlochower, G.M. Green, The limiting oxygen concentration and flammability limits of gases and gas mixtures. J. Loss Prev. Process Ind. **22**(4), 499–505 (2009)
12. E. Goodger, Flammability and ignitability. Appl. Energy **5**(1), 81–84 (1979)
13. T. Ma, S.M. Olenick, M.S. Klassen, R.J. Roby, J.L. Torero, Burning rate of liquid fuel on carpet (porous media). Fire Technol. **40**(3), 227–246 (2004).
14. T. Ma, Q. Wang, M. Larrañaga, From ignition to suppression, a thermal view of flammability limits. Fire Technol. **50**(3), 525–543 (2014)
15. T. Ma, A thermal theory for estimating the flammability limits of a mixture. Fire Saf. J. **46**(8), 558–567 (2011)
16. T. Ma, Using critical flame temperature for estimating lower flammable limits of a mixture. Process Saf. Prog. **32**(4), 387–392 (2013)
17. T. Ma, Q. Wang, M. Larrañaga, Correlations for estimating flammability limits of pure fuels and fuel-inert mixtures. Fire Saf. J. **56**, 9–19 (2013)
18. T. Ma, M. Larrañaga, Theoretical flammability diagram for analyzing mine gases. Fire Technol. **51**(2), 271–286 (2015)
19. T. Ma, M. Larrañaga, Flammability diagrams for oxy-combustion. Fire Technol. (2015) (accepted)
20. T. Ma, A thermal theory for flammability diagrams guiding purge and inertion of a flammable mixture. Process Saf. Prog. **32**(1), 7 (2013)

T. Ma, *Ignitability and Explosibility of Gases and Vapors*,
DOI 10.1007/978-1-4939-2665-7

21. T. Ma, M.D. Larrañaga, Flammable state and dilution requirement in the theoretical flammability diagram. Process Saf. Prog. **33**(1), 70–76 (2014)
22. T. Ma, M.D. Larrañaga, An improved flammability diagram for dilution and purge on gas mixtures. Open J. Chem. Eng. Sci. **1**(1), 18 (2014)
23. P. Daniell, The theory of flame motion. Proc. R. Soc. Lond. Ser. A **126**(802), 393–405 (1930)
24. B. Lewis, G.V. Elbe, Fundamental principles of flammability and ignition. Zeitschrift für Elektrochemie, Berichte der Bunsengesellschaft für physikalische Chemie **61**(5), 574–578 (1957)
25. J. Wehner, On the theory of flammability limits. Combust. Flame **7**, 309–313 (1963)
26. J.O. Hirschfelder, C.F. Curtiss, R.B. Bird, *Molecular Theory of Gases and Liquids*, vol. 26 (Wiley, New York, 1954)
27. M. Burdon, J. Burgoyne, F. Weinberg, The effect of methyl bromide on the combustion of some fuel-air mixtures, in *Symposium (International) on Combustion* (Elsevier, 1955)
28. J. Linnett, C. Simpson, Limits of inflammability, in *Symposium (International) on Combustion* (Elsevier, 1957)
29. D. Spalding, A theory of inflammability limits and flame-quenching. Proc. R. Soc. Lond. A **240**(1220), 83–100 (1957)
30. E. Mayer, A theory of flame propagation limits due to heat loss. Combust. Flame **1**(4), 438–452 (1957)
31. A. Berlad, C. Yang, A theory of flame extinction limits. Combust. Flame **4**, 325–333 (1960)
32. B. Lewis, G. Von Elbe, *Combustion, Flames and Explosions of Gases*, vol. 3 (Cambridge University Press, Cambridge, 1961)
33. C. Law, F. Egolfopoulos, A unified chain-thermal theory of fundamental flammability limits, in *Symposium (International) on Combustion* (Elsevier, 1992)
34. J.S. T'ien, Diffusion flame extinction at small stretch rates: the mechanism of radiative loss. Combust. Flame **65**(1), 31–34 (1986)
35. A. Edgerton, J. Powling, The limits of flame propagation at atmospheric pressure, II. The influences of Changes in the physical properties. Proc. R. Soc. **193A** (1948)
36. H. Le Chatelier, Estimation of firedamp by flammability limits. Ann. Mines **19**(8), 7 (1891)
37. H.W. Frank Coward, C. Carpenter, The dilution limits of inflammability of gaseous mixtures. Part III: The lower limits of some mixed inflammable gases with air. Part IV: The upper limits of some gases, singly and mixed, in air. J. Chem. Soc. Trans. **115**, 27–36 (1919)
38. M. Hertzberg, The theory of flammability limits. Nat. Convection. BuMines RI **8127**, 15 (1976)
39. M. Hertzberg, *The Theory of Flammability Limits. Conductive-Convective Wall Losses and Thermal Quenching* (1980)
40. M. Hertzberg, *The Theory of Flammability Limits. Radiative Losses and Selective Diffusional Demixing* (1982)
41. G. Melhem, A detailed method for estimating mixture flammability limits using chemical equilibrium. Process Saf. Prog. **16**(4), 203–218 (1997)
42. H. Le Chatelier, C. Boudouard, On the limits of inflammability of combustible mixtures. Comptes Rendus des Seances de L' Academie Sciences **126** (1898)
43. M.G. Zabetakis, Flammability characteristics of combustible gases and vapors. DTIC Document (1965)
44. A.G. White, XCVI—Limits for the propagation of flame in inflammable gas–air mixtures. Part III: The effects of temperature on the limits. J. Chem. Soc. Trans. **127**, 672–684 (1925)
45. L.A. Medard, *Accidental Explosions* (Wiley, New York, 1989)
46. A.E. Spakowski, Pressure limits of flame propagation of pure hydrocarbon-air mixtures at reduced pressure (1952)
47. B. Hanley, A model for the calculation and the verification of closed cup flash points for multicomponent mixtures. Process Saf. Prog. **17**(2), 86–97 (1998)
48. L.G. Britton, Using heats of oxidation to evaluate flammability hazards. Process Saf. Prog. **21**(1), 31–54 (2002)

49. M. Zabetakis, S. Lambiris, G. Scott, Flame temperatures of limit mixtures, in *Symposium (International) on Combustion* (Elsevier, 1958)
50. A. Egerton, J. Powling, The limits of flame propagation at atmospheric pressure, I. The influence ofpromoters'. Proc. R. Soc. Lond. A **193**(1033), 172–190 (1948)
51. I. Wierzba, G. Karim, The flammability of fuel mixtures in air containing propane and butane. J. Energy Res. Technol. (USA) **111**(2) (1989)
52. Z. Li et al., Effect of low temperature on the flammability limits of methane/nitrogen mixtures. Energy **36**(9), 5521–5524 (2011)
53. M. Goethals et al., Experimental study of the flammability limits of toluene–air mixtures at elevated pressure and temperature. J. Hazard. Mater. **70**(3), 93–104 (1999)
54. F. Van den Schoor, F. Verplaetsen, The upper explosion limit of lower alkanes and alkenes in air at elevated pressures and temperatures. J. Hazard. Mater. **128**(1), 1–9 (2006)
55. J.E. Hustad, O.K. Sønju, Experimental studies of lower flammability limits of gases and mixtures of gases at elevated temperatures. Combust. Flame **71**(3), 283–294 (1988)
56. G. Ciccarelli, D. Jackson, J. Verreault, Flammability limits of NH_3–H_2–N_2–air mixtures at elevated initial temperatures. Combust. Flame **144**(1), 53–63 (2006)
57. J.R. Rowley, R.L. Rowley, W.V. Wilding, Experimental determination and re-examination of the effect of initial temperature on the lower flammability limit of pure liquids. J. Chem. Eng. Data **55**(9), 3063–3067 (2010)
58. J. Rowley, R. Rowley, W. Wilding, Estimation of the lower flammability limit of organic compounds as a function of temperature. J. Hazard. Mater. **186**(1), 551–557 (2011)
59. L.G. Britton, D.J. Frurip, Further uses of the heat of oxidation in chemical hazard assessment. Process Saf. Prog. **22**(1), 1–19 (2003)
60. L. Catoire, V. Naudet, Estimation of temperature-dependent lower flammability limit of pure organic compounds in air at atmospheric pressure. Process Saf. Prog. **24**(2), 130–137 (2005)
61. S. Kondo et al., On the temperature dependence of flammability limits of gases. J. Hazard. Mater. **187**(1), 585–590 (2011)
62. A. Shimy, Calculating flammability characteristics of hydrocarbons and alcohols. Fire Technol. **6**(2), 135–139 (1970)
63. P. Lloyd, The fuel problem in gas turbines. Inst. Mech. Eng. Proc. War Emergency **159**(41) (1948)
64. C. Beyler, *Flammability Limits of Premixed and Diffusion Flames.* SFPE Handbook of Fire Protection Engineering, vol. 3 (2002)
65. A. Spakowski, A. Belles, E. Frank, Variation of pressure limits of flame propagation with tube diameter for various isooctane-oxygen-nitrogen mixtures (1952)
66. C.V. Mashuga, D.A. Crowl, Derivation of Le Chatelier's mixing rule for flammable limits. Process Saf. Prog. **19**(2), 112–117 (2000)
67. J. Yeaw, Explosive limits of industrial gases. Indus. Eng. Chem. **21**(11), 1030–1033 (1929)
68. H.F. Coward, G.W. Jones, Limits of flammability of gases and vapors. DTIC Document (1952)
69. W. Heffington, W. Gaines, Flammability calculations for gas mixtures. Oil Gas J. **79**, 87 (1981)
70. M. Molnarne, V. Schröder, Comparison of calculated data for the flammability and the oxidation potential according to ISO 10156 with experimentally determined values. J. Loss Prev. Process Ind. **24**(6), 900–907 (2011)
71. commitee, I.t., ISO 10156:2010 (E) Gases and gas mixtures—Determination of fire potential and oxidizing ability for the selection of cylinder valve outlets (2010)
72. G.W. Jones, R.E. Kennedy, I. Spolan, *Effect of High Pressures on the Flammability of Natural Gas–Air–Nitrogen Mixtures* (Bureau of Mines, Pittsburgh, 1949)
73. A.E. Spakowski, Pressure limits of flame propagation of pure hydrocarbon-air mixtures at reduced pressure. National Advisory Committee for Aeronautics (1952)
74. F.Y. Hshieh, Predicting heats of combustion and lower flammability limits of organosilicon compounds. Fire Mater. **23**(2), 79–89 (1999)

75. T. Suzuki, Note: Empirical relationship between lower flammability limits and standard enthalpies of combustion of organic compounds. Fire Mater. **18**(5), 333–336 (1994)
76. A. Pintar, Predicting lower and upper flammability limits, in *Proceedings of the International Conference on Fire Safety* (1999)
77. C. Hilado, A method for estimating limits of flammability (organic compound hazards). J. Fire Flamm. **6**, 130–139 (1975)
78. V.T. Monakhov, *Methods for Studying the Flammability of Substances*. (Amerind Publishing Company, Hardcover, 1986)
79. C. Beyler, Flammability limits of premixed and diffuon flames, in *SFPE Handbook of Fire Protection Engineering*, ed. by P. DiNenno (NFPA publishing, Quincy, 2008), pp. 2–194
80. A.B. Donaldson, N. Yilmaz, A. Shouman, Correlation of the flammability limits of hydrocarbons with the equivalence ratio. Int. J. Appl. Eng. Res **1**(1), 77–85 (2006)
81. C.J. Hilado, H.J. Cumming, The HC value: a method for estimating the flammability of mixtures of combustible gases. Fire Technol. **13**(3), 195–198 (1977)
82. P.J. DiNenno, *Halon Replacement Clean Agent Total Flooding Systems*. The SFPE Handbook of Fire Protection Engineering, p. 7 (2002)
83. G.W. Jones, *Inflammability of Mixed Gases*, vol. 450 (US Govt. print. off, 1929)
84. W. Haessler, *Fire Fundamentals and Control* (1990)
85. J. Cheng, C. Wang, S. Zhang, Methods to determine the mine gas explosibility—an overview. J. Loss Prev. Process Ind. **25**(3), 425–435 (2012)
86. D.A. Crowl, *Understanding Explosions*, vol. 16 (Wiley-AIChE, New York, 2010)
87. V. Babrauskas, *Ignition Handbook*, vol. 1 (Fire Science Publishers Issaquah, 2003)
88. F. Mowrer, Fundamentals of the Fire Hazards of Materials, in *Handbook of Building Materials for Fire Protection,* ed. by C.A. Harper (McGraw-Hill, New York, 2004)
89. W. Thornton, *XV.* The relation of oxygen to the heat of combustion of organic compounds. Lond. Edinb. Dublin Philos. Mag. J. Sci. **33**(194), 196–203 (1917)
90. C. Huggett, Estimation of rate of heat release by means of oxygen consumption measurements. Fire Mater. **4**(2), 61–65 (1978)
91. J.G. Quintiere, *Fundamentals of Fire Phenomena* (Wiley, Chichester, 2006)
92. K.K. Kuo, *Principles of Combustion* (1986)
93. D. Drysdale, *An Introduction to Fire Dynamics* (Wiley, New York, 1998)
94. D. Drysdale, *An Introduction to Fire Dynamics* (Wiley.com, New York, 2011)
95. A.D. Putorti, J.A. McElroy, D. Madrzykowski, *Flammable and Combustible Liquid Spill/Burn Patterns* (US Department of Justice, Office of Justice Programs, National Institute of Justice, 2001)
96. N.F.P. Association, N.F.P. Association, *NFPA 30: Flammable and Combustible Liquids Code* (National Fire Protection Association, Quincy, 2011)
97. D.A. Crowl, J.F. Louvar, *Chemical Process Safety: Fundamentals With Applications* (Prentice Hall, Englewood Cliffs, 2001)
98. A. Kanury, Combustion characteristics of biomass fuels. Combust. Sci. Technol. **97**(4–6), 469–491 (1994)
99. T. Suzuki, M. Ishida, Neural network techniques applied to predict flammability limits of organic compounds. Fire Mater. **19**(4), 179–189 (1995)
100. A.W. Akhras, M.L. Wagner, *Method for Predicting Flammability Limits of Complex Mixtures* (Google Patents, 2002)
101. Anon, *NIST Chemistry Webbook* (2013). http://webbook.nist.gov/chemistry/
102. T. Ma, Using critical flame temperature for estimating lower flammable limits of a mixture. Process Saf. Prog. **32**(4), 6 (2013)
103. Y.N. Shebeko et al., An analytical evaluation of flammability limits of gaseous mixtures of combustible–oxidizer–diluent. Fire Saf. J. **37**(6), 549–568 (2002)
104. A. Maček, Flammability limits: a re-examination. Combust. Sci. Technol. **21**(1–2), 43–52 (1979)

105. Y. Zhao, L. Bin, Z. Haibo, Experimental study of the inert effect of R134a and R227ea on explosion limits of the flammable refrigerants. Exp. Thermal Fluid Sci. **28**(6), 557–563 (2004)
106. S. Kondo et al., A study on flammability limits of fuel mixtures. J. Hazard. Mater. **155**(3), 440–448 (2008)
107. G. Jones, R. Kennedy, Prevention of gas explosions by controlling oxygen concentration. Indus. Eng. Chem. **27**(11), 1344–1346 (1935)
108. D. Razus, M. Molnarne, O. Fuß, Limiting oxygen concentration evaluation in flammable gaseous mixtures by means of calculated adiabatic flame temperatures. Chem. Eng. Process. **43**(6), 775–784 (2004)
109. C.V. Mashuga, D.A. Crowl, Application of the flammability diagram for evaluation of fire and explosion hazards of flammable vapors. Process Saf. Prog. **17**(3), 176–183 (1998)
110. F. Girodroux, A. Kusmierz, C.J. Dahn, Determination of the critical flammability ratio (CFR) of refrigerant blends. J. Loss Prev. Process Ind. **13**(3), 385–392 (2000)
111. D.P. Wilson, R.G. Richard, Determination of refrigerant lower flammability limits in compliance with proposed addendum p to Standard 34 (RP-1073). Trans. Am. Soc. Heat. Refrig. Air. Cond. Eng, **108**(2), 739−756 (2002).
112. H.W. Carhart, W.A. Affens, Flammability of aircraft fuels. Fire Technol. **5**(1), 16–24 (1969)
113. S. Ishizuka, H. Tsuji, An experimental study of effect of inert gases on extinction of laminar diffusion flames, in *Symposium (International) on Combustion* (Elsevier, New York, 1981)
114. R. Simmons, H. Wolfhard, Some limiting oxygen concentrations for diffusion flames in air diluted with nitrogen. Combust. Flame **1**(2), 155–161 (1957)
115. H. Persson, A. Lönnermark, Tank fires. SP report, vol. 14 (2004)
116. E. Planas-Cuchi, J.A. Vílchez, J. Casal, Fire and explosion hazards during filling/emptying of tanks. J. Loss Prev. Process Ind. **12**(6), 479–483 (1999)
117. J. Chomiak, J. Longwell, A. Sarofim, Combustion of low calorific value gases; problems and prospects. Prog. Energy Combust. Sci. **15**(2), 109–129 (1989)
118. H. Tsuji et al., *High Temperature Air Combustion: From Energy Conservation to Pollution Reduction* (CRC Press, Boca Raton, 2010)
119. L. Muniz, M. Mungal, Instantaneous flame-stabilization velocities in lifted-jet diffusion flames. Combust. Flame **111**(1), 16–31 (1997)
120. G.W. Jones, R. E. Kennedy, *Limits of Inflammability of Natural Gases Containing High Percentages of Carbon Dioxide and Nitrogen* (1933)
121. L. Chen, S.Z. Yong, A.F. Ghoniem, Oxy-fuel combustion of pulverized coal: characterization, fundamentals, stabilization and CFD modeling. Prog. Energy Combust. Sci. **38**(2), 156–214 (2012)
122. D.S. Burgess et al., *Flammability of Mixed Gases* (Bureau of Mines, Pittsburgh, 1982)
123. W.M. Heffington, W.R. Gaines, D.A. Renfroe, Flammability limits of coal-derived low-btu gas mixtures containing large amounts of inert gases. Combust. Sci. Technol. **36**(3–4), 191–197 (1984)
124. G. Jones, The flammability of refrigerants. Ind. Eng. Chem. **20**(4), 367−370 (1928)
125. B. Dlugogorski, R. Hichens, E. Kennedy, Inert hydrocarbon-based refrigerants. Fire Saf. J. **37**(1), 53–65 (2002)
126. Z. Yang, Y. Li, Q. Zhu, et al. Research on non-flammable criteria on refrigerants. Appl. Therm. Eng. **20**(14):1315−1320 (2000)
127. V. Schröder, M. Molnarne, Flammability of gas mixtures. Part 1: fire potential. J. Hazard. Mater. **121**(1), 37–44 (2005)
128. M. Kukuczka, A new method for determining explosibility of complex gas mixtures. Mechanizacja I Automatuzacja Gornictwa **164**(11), 36–39 (1982)
129. S.H. Ash, E.W. Felegy, *Analysis of Complex Mixtures of Gases*, ed. by B.o. Mines (1948)
130. I. Shaw*A Graphic Method Of Determinin G Ti-*-Explos Ilit Y Characteristics of Mine-Fire Atmospi-*-Res.*

131. R.J. Timko, R. Derick, *Methods to Determine the Status of Mine Atmospheres—an Overview* (National Institute for Occupational Safety and Health Publications, 2006)
132. R.E. Greuer, *Study of Mine Fire Fighting Using Inert Gases* (Michigan Technological University, College of Engineering, Department of Mining Engineering, USA, 1974)
133. J. Cheng, Y. Luo, Modified explosive diagram for determining gas-mixture explosibility. J. Loss Prev. Process Ind. (2013)
134. D. Cray, *Deep Underground, Miles of Hidden Wildfires Rage* (2011) (cited 23 July 2010). http://www.time.com/time/health/article/0,8599,2006195,00.html
135. G.B. Stracher, A. Prakash, E.V. Sokol, *Coal and Peat Fires: A Global Perspective: Volume 1: Coal-Geology and Combustion*. vol. 1 (Elsevier, Amsterdam, 2010)
136. G.B. Stracher, T.P. Taylor, Coal fires burning out of control around the world: thermodynamic recipe for environmental catastrophe. Int. J. Coal Geol. **59**(1), 7–17 (2004)
137. R. Morris, T. Atkinson, Sampling gases in a sealed mine fire area. Min. Sci. Technol. **5**(1), 25–31 (1987)
138. J. Graham, The origin of blackdamp. Trans. Inst. Min. Eng. **55**, 1917–1918 (1917)
139. J. Jones, J. Trickett, Some observations on the examination of gases resulting from explosions in colleries. Trans. Inst. Min. Eng. **44**, 768–791 (1955)
140. D.W. Mitchell, *Mine Fires* (Maclean Hunter Publications, Chicago, 1990)
141. C. Litton, A. Kim, Improved mine fire diagnostic techniques. Soc. Min. Metall. Explor. Annu. Mtg. SME Prepr. **89–182**, 7 (1989)
142. C.D. Litton, *Gas Equilibrium in Sealed Coal Mines* (US Department of the Interior, Bureau of Mines, 1986)
143. T.R. Justin, A.G. Kim, Mine fire diagnostics to locate and monitor abandoned mine fires. BuMines IC **9184**, 348–355 (1988)
144. M.G. Zabetakis, R.W. Stahl, H.A. Watson, *Determining the Explosibility of Mine Atmospheres*, vol. 7901 (US Department of the Interior, Bureau of Mines, 1959)
145. J. Cheng, Y. Luo, F. Zhou, Explosibility safety factor: an approach to assess mine gas explosion risk. Fire Technol. 1–15 (2013)
146. S. Ray et al., Assessing the status of sealed fire in underground coal mines. J. Sci. Ind. Res. **63**(7), 579–591 (2004)
147. A.K. Singh et al., Mine fire gas indices and their application to Indian underground coal mine fires. Int. J. Coal Geol. **69**(3), 192–204 (2007)
148. T. Ma, M.D. Larrañaga, Theoretical flammability diagrams for diagnosing mine gases. Fire Technol. (2013) (Accepted)
149. L.-G. Bengtsson, *Enclosure Fires* (Swedish Rescue Services Agency, Karlstad, 2001)
150. S. Kerber, D. Madrzykowski, Fire fighting tactics under wind driven fire conditions: 7-story building experiments. NIST technical note, vol. 1629 (2009)
151. D. Madrzykowski, S.I. Kerber, *Fire Fighting Tactics Under Wind Driven Conditions: Laboratory Experiments* (Fire Protection Research Foundation, Quincy, 2009)
152. W. Croft, Fires involving explosions—a literature review. Fire Saf. J. **3**(1), 3–24 (1980)
153. C. Fleischmann, P. Pagni, R. Williamson, Quantitative backdraft experiments, in *1993 Annual Conference on Fire Research: Book of Abstracts. NISTIR* (1994)
154. D.T. Gottuk et al., The development and mitigation of backdraft: a real-scale shipboard study. Fire Saf. J. **33**(4), 261–282 (1999)
155. W. Weng, W. Fan, Critical condition of backdraft in compartment fires: a reduced-scale experimental study. J. Loss Prev. Process Ind. **16**(1), 19–26 (2003)
156. J. Mao,Y.H. Xi, G. Bai, H.M. Fan, H.Z Ji, A model experimental study on backdraught in tunnel fires. Fire Saf. J. **46**(4), 164–177 (2011)
157. A. Chen et al., Theoretical analysis and experimental study on critical conditions of backdraft. J. Loss Prev. Process Ind. **24**(5), 632–637 (2011)
158. H.F. Coward, F. Brinsley, CLXXIV—The dilution-limits of inflammability of gaseous mixtures. Part I: The determination of dilution-limits. Part II: the lower limits for hydrogen, methane, and carbon monoxide in air. J. Chem. Soc. Trans. **105**, 1859–1885 (1914)

159. H.B. Dixon, J.M. Crofts, CXC—The firing of gases by adiabatic compression. Part II: the ignition-points of mixtures containing electrolytic gas. J. Chem. Soc. Trans. **105**, 2036–2053 (1914)
160. H. Tizard, D. Pye, *VIII.* Experiments on the ignition of gases by sudden compression. Lond. Edinb. Dublin Philos. Mag. J. Sci. **44**(259), 79–121 (1922)
161. H.B. Dixon, H.F. Coward, LXVII.—The ignition-temperatures of gases. J. Chem. Soc. Trans. **95**, 514–543 (1909)
162. C.V. Mashuga, D.A. Crowl, Flammability zone prediction using calculated adiabatic flame temperatures. Process Saf. Prog. **18**(3), 127–134 (1999)
163. M. Hertzberg, *Theory of Flammability Limits. Flow Gradient Effects and Flame Stretch* (1984)
164. C. Beyler, A brief history of the prediction of flame extinction based upon flame temperature. Fire Mater. **29**(6), 425–427 (2005)
165. M.R. Brooks, D.A. Crowl, Flammability envelopes for methanol, ethanol, acetonitrile and toluene. J. Loss Prev. Process Ind. **20**(2), 144–150 (2007)
166. D. Mathieu, Power law expressions for predicting lower and upper flammability limit temperatures. Indus. Eng. Chem. Res. **52**(26), 9317–9322 (2013)
167. C.V. Mashuga, D.A. Crowl, Flammability zone prediction using calculated adiabatic flame temperatures. Process Saf. Prog. **18**(3), 127–134 (2004)
168. M. Molnarne, P. Mizsey, V. Schröder, Flammability of gas mixtures. Part 2: influence of inert gases. J. Hazard. Mater. **121**(1), 45–49 (2005)
169. V. Schroeder, K. Holtappels, Explosion characteristics of hydrogen-air and hydrogen-oxygen mixtures at elevated pressures, in *International Conference on Hydrogen Safety,* Congress Palace, Pisa, Italy 2005
170. J.G. Hansel, J.W. Mitchell, H.C. Klotz, Predicting and controlling flammability of multiple fuel and multiple inert mixtures. Plant/Oper. Prog. **11**(4), 213–217 (1992)
171. Y. Ju, G. Masuya, P.D. Ronney, Effects of radiative emission and absorption on the propagation and extinction of premixed gas flames, in *Symposium (international) on Combustion* (Elsevier, 1998)
172. Z. Chen et al., Studies of radiation absorption on flame speed and flammability limit of CO_2 diluted methane flames at elevated pressures. Proc. Combust. Inst. **31**(2), 2693–2700 (2007)

Printed in the United States
By Bookmasters